Responsible Rural Tourism in Asia

ASPECTS OF TOURISM

Series Editors: Chris Cooper (*Leeds Beckett University, UK*), C. Michael Hall (*University of Canterbury, New Zealand*) and Dallen J. Timothy (*Arizona State University, USA*)

Aspects of Tourism is an innovative, multifaceted series, which comprises authoritative reference handbooks on global tourism regions, research volumes, texts and monographs. It is designed to provide readers with the latest thinking on tourism worldwide and in so doing will push back the frontiers of tourism knowledge. The series also introduces a new generation of international tourism authors writing on leading edge topics.

The volumes are authoritative, readable and user-friendly, providing accessible sources for further research. Books in the series are commissioned to probe the relationship between tourism and cognate subject areas such as strategy, development, retailing, sport and environmental studies. The publisher and series editors welcome proposals from writers with projects on the above topics.

All books in this series are externally peer-reviewed.

Full details of all the books in this series and of all our other publications can be found on http://www.channelviewpublications.com, or by writing to Channel View Publications, St Nicholas House, 31–34 High Street, Bristol BS1 2AW, UK.

ASPECTS OF TOURISM: 89

Responsible Rural Tourism in Asia

Edited by
**Vikneswaran Nair,
Amran Hamzah and Ghazali Musa**

CHANNEL VIEW PUBLICATIONS
Bristol • Blue Ridge Summit

DOI https://doi.org/10.21832/NAIR7512
Library of Congress Cataloging in Publication Data
A catalog record for this book is available from the Library of Congress.
Names: Nair, Vikneswaran- editor. | Amran, Hamzah, editor. | Musa, Ghazali, editor.
Title: Responsible Rural Tourism in Asia/Edited by Vikneswaran Nair, Amran Hamzah and Ghazali Musa.
Description: Bristol; Blue Ridge Summit: Channel View Publications, [2020] | Series: Aspects of Tourism: 89 | Includes bibliographical references and index. | Summary: "Rural tourism has contributed to the income generation of countries in Asia thereby reducing poverty and improving quality of life. This book explores the fundamentals of responsible rural tourism. It covers a range of Asian countries and examines both successful and failed attempts in developing responsible rural tourism"— Provided by publisher.
Identifiers: LCCN 2020019020 (print) | LCCN 2020019021 (ebook) | ISBN 9781845417512 (hardback) | ISBN 9781845417529 (pdf) | ISBN 9781845417536 (epub) | ISBN 9781845417543 (kindle edition)
Subjects: LCSH: Rural tourism—Asia. | Rural tourism—Social aspects—Asia. | Sustainable tourism—Asia.
Classification: LCC G155.A78 R47 2020 (print) | LCC G155.A78 (ebook) | DDC 338.4/791504—dc23 LC record available at https://lccn.loc.gov/2020019020
LC ebook record available at https://lccn.loc.gov/2020019021

British Library Cataloguing in Publication Data
catalogue entry for this book is available from the British Library.

ISBN-13: 978-1-84541-751-2 (hbk)
ISBN-13: 978-1-84541-117-6 (pbk)

Channel View Publications
UK: St Nicholas House, 31–34 High Street, Bristol, BS1 2AW, UK.
USA: NBN, Blue Ridge Summit, PA, USA.

Website: www.channelviewpublications.com
Twitter: Channel_View
Facebook: https://www.facebook.com/channelviewpublications
Blog: www.channelviewpublications.wordpress.com

Copyright © 2020 Vikneswaran Nair, Amran Hamzah, Ghazali Musa and the authors of individual chapters.

All rights reserved. No part of this work may be reproduced in any form or by any means without permission in writing from the publisher.

The policy of Multilingual Matters/Channel View Publications is to use papers that are natural, renewable and recyclable products, made from wood grown in sustainable forests. In the manufacturing process of our books, and to further support our policy, preference is given to printers that have FSC and PEFC Chain of Custody certification. The FSC and/or PEFC logos will appear on those books where full certification has been granted to the printer concerned.

Typeset by Nova Techset Private Limited, Bengaluru and Chennai, India.

Contents

	Contributors	vii
	Acknowledgements	xiii

Introduction

1 Conceptualizing Responsible Rural Tourism in Asia 1
   Vikneswaran Nair, Ghazali Musa and Amran Hamzah

Theme 1: Harmony Between Humans and Nature in Moulding Rural Tourism in Asia

2 Responsible Rural Tourism in Bhutan: Aligning Gross National Happiness with the Cape Town Principles 27
   Sarah Schiffling, Chris Phelan and Karma Pema Loday

3 Crown Cave, Guilin: A Chinese Perspective on Responsible Rural Tourism 41
   Trevor H.B. Sofield and Fung Mei Sarah Li

Theme 2: An Essentially Asian Form of Rural Tourism

4 Responsible Rural Tourism in Japan's Tea Villages 61
   Lee Jolliffe and Moe Nakashima

5 Community Garden Experience During the Off-Peak Tourism Rainy Season at Hoi An Heritage Town, Vietnam 75
   Thu Thi Trinh

6 Responsible Rural Tourism Initiatives and Local Community Development in Kerala, India 87
   Anu Treesa George, Terry DeLacy and Min Jiang

Theme 3: Tourism That Takes Place in Rural Areas

7 Linking Responsible Rural Tourism to Agritourism in the Philippines: A Case Study of Costales Nature Farms 103
   Miguela M. Mena and Charmielyn C. Sy

8   Is Community-Based Tourism a Tool for the Sustainability
    of the Local Community and the Local Economy? The Case
    of Coruh Valley, Turkey   116
    Sıla Karacaoğlu and Medet Yolal

9   Community Characteristics, Social Cohesion and the Success
    of Community-Based Tourism: Case Studies of Vietnam   128
    Tramy Ngo and Nguyen Thi Huyen

10 Conversations with the Local Champions of
    Miso Walai Homestay: Responsible Tourism in Practice   141
    Amran Hamzah

11 'MlupBaitong' – A Pioneer in Responsible Rural Tourism in
    Cambodia   162
    Trevor H.B. Sofield

12 Can Tourism be a Success Story? Stakeholders' Management
    and Narratives of Rural Tourism. Reflexive Analysis of a
    Tourism Project in Timor Leste   177
    Frederic Bouchon

13 The Constructs of Responsible Rural Tourism Governance
    for Belum-Temengor Forest Reserve, Malaysia   195
    Joo-Ee Gan and Vikneswaran Nair

Conclusion

14 The Quest for an Essentially Asian Form of Rural Tourism   213
    Amran Hamzah, Vikneswaran Nair and Ghazali Musa

    Index   223

Contributors

Editors

Vikneswaran Nair is a Professor in Sustainable Tourism and the Dean of Graduate Studies and Research at University of The Bahamas (UB). Prior to joining UB in 2017, he was at Taylor's University, Malaysia, for 19 years as a full Professor of Sustainable Tourism, Programme Leader for the Responsible Rural Tourism Network; Research Fellow of the Centre for Research and Innovation in Tourism (CRiT) and the founding Director of the Centre for Research and Development at the University. A consultant with many national and international projects, his exceptional research achievements and publications, have earned him many international and national awards. His research specialization is in Sustainable and Responsible Tourism, Rural Tourism, Ecotourism Management, Environmental Management, Community-based Tourism and Green Tourism.

Ghazali Musa is a medical doctor (MBBS, Malaya) and has a PhD in tourism from Otago University (New Zealand). He practiced medicine for five years in Malaysia and Singapore before embarking in tourism academia. His main research interests are mountaineering tourism, scuba diving tourism and health tourism. Ghazali is currently a full professor in tourism management (since 2012) and the Head of Business Strategy and Policy Department, Faculty of Business and Accountancy, University Malaya (Malaysia). He has produced over 100 publications which include 10 books, 30 book chapters and 50 refereed journal articles. He is on the editorial boards of several international tourism journals.

Amran Hamzah is a Professor in Tourism Planning and Director of the Centre for Innovative Planning and Development (CIPD) at Universiti Teknologi Malaysia. He is an academic practitioner who specializes in tourism policy planning and the interface between tourism and conservation. Besides his academic duties, he has been active as a tourism consultant, having led more than 70 consultancy projects for international agencies such as APEC, the ASEAN Secretariat, IUCN and UNESCO as well as national agencies, notably the Ministry of Tourism, Arts and Culture Malaysia (MOTAC) and Tourism Malaysia. Amran was the lead consultant for the National Ecotourism Plan (2016–2025) and the National Tourism Policy (2020–2030) which were commissioned by MOTAC.

Authors

Sarah Schiffling is a Senior Lecturer in Supply Chain Management at Liverpool John Moores University, UK, and an International Research Fellow with the HUMLOG Institute in Helsinki, Finland. She holds a PhD from Heriot-Watt University, UK, focusing on complexity in humanitarian logistics. Her research interests include humanitarian logistics, coopetition, and supply chain management in developing nations. She is particularly interested in the interactions of supply chains with communities.

Chris Phelan is a Senior Lecturer in Business and Management at Edge Hill University, UK. He completed his PhD at the University of Central Lancashire, UK, which considered the entrepreneurial and managerial competencies associated with farm diversification and agritourism. His ongoing research continues to consider the contribution of entrepreneurs and small business owners to local economic development, particularly the role of competencies and skills development within regional entrepreneurial ecosystems.

Karma Pema Loday is an Assistant Professor in Management Studies at the Royal Institute of Management, Simtokha, Bhutan. He has been working at the institute since October 1996. Currently, he is the Head of the Department of Finance and Business. He has worked extensively as an organizational development consultant for various government and corporate organizations. His interests in research and consultancy are in the areas of entrepreneurship, organizational development and change management.

Trevor H.B. Sofield is retired Foundation Professor of Tourism, University of Tasmania; and Visiting Professor, Chinese University of Hong Kong, Sun Yat Sen University, China, and University of Girona, Spain. He is internationally active in research, consultancies and teaching, with 400+ publications in tourism policy, planning and development across a range of areas at national, regional and local levels. Interests are eclectic and include community-based tourism, ecotourism, heritage tourism, protected area management, wildlife tourism, politics and governance, value chain analysis, tourism education, and indigenous tourism, with Asia and the South Pacific as the main geographical regions of focus.

Fung Mei Sarah Li has degrees from the Chinese University of Hong Kong, University of Surrey, England and Murdoch University, Perth, Western

Australia. Recently retired from full-time teaching, she continues to be active in undertaking tourism research and consultancies in China, East Asia and the South Pacific and as a visiting professor with the University of Girona, Spain. Her academic record across three continents finds expression in her ability to bridge Asian and western values in areas such as policy and planning for tourism development and poverty alleviation, cultural tourism, the geography of tourism and value chain analysis.

Lee Jolliffe is Professor (Retired) at the University of New Brunswick, Saint John, Canada and a Visiting Professor at Asia Ritsumeikan University, Japan. She is the editor of *Tea and Tourism: Tourists, Traditions and Transformations* (2007) published with Channel View Publications and has investigated tea tourism in Japan, China, Thailand, Sri Lanka, Taiwan and South Korea.

Moe Nakashima is Internship Co-ordinator with Kyoto Obobu Tea Farms in Wazuka, Japan. She is a graduate of Kochi University, Japan. During her undergraduate studies in agriculture she spent a term at the University of New Brunswick, Saint John, Canada studying tea tourism with Professor Lee Jolliffe.

Thu Thi Trinh is Director of Center for Economic Studies, Research Institute of Central Region, Vietnam Academy of Social Sciences (VASS). She gained her PhD in Tourism Management from The University of Waikato, New Zealand. Her research interests lie in museums, culture and heritage as tourist attractions. Her research papers have been published in *Tourism Management, Journal of Travel Research, Current Issues in Tourism* and *Journal of Vacation Marketing*. She is the Vietnam country representative of Asia Pacific CHRIE and was Vice Dean of Research of Danang College of Economics-Planning and Principal of Pegasus International College.

Anu Treesa George is a PhD candidate at the Research Institute of Sustainable Industries and Liveable Cities at Victoria University, Australia. Her research interest is in responsible tourism, hospitality and events industries, green growth strategies and community well-being. Her research focuses on mitigating the sustainability challenges by developing a practical framework which combines the green growth concepts with responsible tourism to enhance the sustainable practices in tourism industry.

Terry DeLacy is a Professor in Sustainable Tourism in the Research Institute of Sustainable Industries and Liveable Cities, Victoria University, Australia. Previously DeLacy was Director of the Australian National Sustainable Tourism Cooperative Research Centre and Professor of

Environmental Policy at the University of Queensland. His research interests are in natural resources and sustainable tourism, most recently focusing on transformation of the tourism sector into the emerging green economy.

Min Jiang is a senior research fellow at the School of Geography, the University of Melbourne, and a higher degree research supervisor at Research Institute of Sustainable Industries and Liveable Cities, Victoria University. Her current research examines the interactions between infrastructure construction and market-oriented water governance mechanisms in the context of China's water law and policy reform. Holding a PhD in Environmental Law, Dr Jiang conducts interdisciplinary, policy-oriented research across water governance, climate change adaptation, and destination green growth, with a focus on the Asia Pacific region. Dr Jiang is the sole author of the book *Towards Tradable Water Rights: Water Law and Policy Reform in China* (Springer, 2018).

Miguela M. Mena is a Professor and former Dean of the University of the Philippines Asian Institute of Tourism. She obtained her PhD in Tourism Management at The Hong Kong Polytechnic University School of Hotel and Tourism Management, Master of Statistics and Bachelor of Science in Statistics at the University of the Philippines School of Statistics, and Diploma for Tourism Management (Highest Distinction) from Institute of Tourism and Hotel Management in Salzburg, Austria.

Charmielyn C. Sy. is an Assistant Professor at the University of the Philippines Asian Institute of Tourism. She obtained her Master of Arts in Philippine Studies at the University of the Philippines Asian Center and her Bachelor of Science in Tourism at the University of the Philippines Asian Institute of Tourism. She is also the Managing Director of Pacific Voyage Travel and Tours, Inc. and a certified ASEAN Master Trainer.

Sıla Karacaoğlu, PhD is an Assistant Professor in the Bilecik Şeyh Edebali University, Faculty of Applied Sciences, department of Tour Guiding. She received her PhD degree from the Mersin University in Tourism Management in 2017. In her doctoral dissertation, she investigates 'Community Perceptions, Attitudes and Support for Community Based Tourism: The Case of Misi Village'. Her research interests are tourism marketing, sustainable tourism and its types, special interest tourism and cultural heritage.

Medet Yolal, PhD is Professor of Marketing in Faculty of Tourism at Anadolu University, Turkey, where he mainly teaches issues related to destination management and marketing, tourism marketing and consumer behaviour. He has authored or co-authored several articles, book chapters and conference papers on hospitality marketing, consumer behaviour, management of small and medium sized enterprises in tourism, and event management. His research interests mainly focus on tourism marketing, consumer behaviour, tourist experience, event management, tourism development and quality of life research in tourism.

Tramy Ngo has a PhD in Tourism Management from Griffith University, Australia, and is currently a researcher of Griffith Institute for Tourism, Griffith University. Her research interests include sustainable tourism, tourism impact assessment, indigenous tourism and cross-cultural studies. She has extensive working experience in tourism research, consultation and tourism higher education. Her works have been published in a number of journals, including *Journal of Sustainable Tourism and Current Issues in Tourism*.

Nguyen Thi Huyen works for the International Labour Organisation (ILO) Viet Nam as a specialist in pro-poor tourism value chain, managing tourism programmes and supporting the formalization of informal economy. Before joining ILO Viet Nam, she worked for different INGOs specialised in improving livelihoods for vulnerable groups. She is also a co-founder of the Viet Nam Community-based Tourism Network. She received her Master's degree in Development Studies from Graduate Institute of International and Development Studies, Geneva, Switzerland. She has published articles and books from her consultation works, including 'The guide for facilitators in community-based tourism development'.

Frederic Bouchon has a background in geography. The theme of his PhD was on the Metropolisation of Kuala Lumpur and its relationship with Tourism. He developed an international expertise before joining Taylor's University, Malaysia in 2006 until 2018. Frederic is a founding member of the ASEAN Tourism Research Association (ATRA), a network committed to the development of Tourism research in the region. His research focuses on Urban Tourism, Tourism Stakeholders, and Destination Planning. Frederic has been involved in Consultancy projects in Malaysia (MICE), Vietnam (Tourism Planning), Timor Leste (Rural Tourism), and for the ASEAN (Road Mobility Tourism).

Joo-Ee Gan is the Director of Undergraduate Studies at the School of Business, Monash University Malaysia. Her research is socio-legal in nature. Focusing on employment in the hotel sector and sustainable practices in the hotel and tourism industry, her studies evaluate the impact of legal measures. Her research on governance examines tourism as a vehicle for protecting the environment and advancing the socioeconomic well-being of local communities, particularly the indigenous people. Prior to joining academia, Joo-Ee was a solicitor specializing in corporate matters. She was also a legal editor of the *Hong Kong Law Report & Digests*. She is an accredited mediator of the Malaysian Bar Council and is on the Panel of the Malaysian Mediation Centre.

Acknowledgements

Our heartfelt appreciation goes to all the chapter authors and co-authors for their contribution to this book and for their patience in making the required changes to complete their task. We would also like to thank all institutions and organizations that supported the authors in completing their chapters. Thank you to the serial reviewers who provided us invaluable feedback to improve each chapter so that it meets the goal of the book.

Special thanks are also rendered to University of The Bahamas for providing the support in getting this book completed. We would also like to extend our gratitude to Mrs Sumangala Pillai of Malaysia for editing and proofreading the chapter manuscripts, despite the tight timeline to complete the task.

This book would also not be possible without the initial work done to develop responsible rural tourism via the Ministry of Higher Education of Malaysia's Long Term Research Grant Scheme 2011 [LRGS grant no: JPT.S (BPKI)2000/09/01/015Jld.4(67)] that was led by Taylor's University, Malaysia from 2011–2015.

Last but not least, we would like to thank the team from Channel View Publications, led by Ms Sarah Williams, for managing the whole process of publication so efficiently.

Professor Vikneswaran Nair, University of The Bahamas
Professor Ghazali Musa, University of Malaya
Professor Amran Hamzah, Universiti Teknologi Malaysia

1 Conceptualizing Responsible Rural Tourism in Asia

Vikneswaran Nair, Ghazali Musa and Amran Hamzah

Introduction

The United Nations (UN) General Assembly declared 2017 as the International Year of Sustainable Tourism. This was a clear recognition of tourism's contribution to sustainable development goals and its potential for community development, poverty eradication and the protection of biodiversity. Tourism is an important tool for community development, especially in rural destinations (Nair *et al.*, 2014, 2017).

Tourism growth and the uniqueness of tourism in Asia have been discussed by many scholars in many case studies over the past few decades (Hall & Page, 2017; Singh, 2011; Winter *et al.*, 2008). As described by Hall and Page (2017: blurb):

> Asia is regarded as the fastest growing area for international and domestic tourism in the world today and over the next 20 years. Given the economic, social and environmental importance of tourism in the region, there is a need for a comprehensive and readable overview of the critical debates and controversies in tourism in the region and the major factors that are affecting tourism development, both now and in the foreseeable future.

Singh (2011) viewed that the many interfaces of the Asian cultural and natural heritage with tourism, while taking into consideration the realities of the current political and economic realities, make the tourism setting in Asia different and important compared with other parts of the world.

Hence, the rural tourism sector has become a key driver for social and economic growth and a major source of income for developing and low-income countries in significant parts of Asia. Rural tourism is progressively viewed as a solution to increasing the economic viability of marginalized areas, stimulating social regeneration and improving the living conditions of rural communities as discussed in the chapters in this book. Similarly, responsible tourism in essence provides quality travel experience that promotes conservation of the natural environment, protects the authenticity of culture and offers socioeconomic opportunities and benefits for local communities. Thus, responsible rural tourism is certainly the way forward for many developing and low-income economies, especially in Asia.

After almost two decades of bringing rural tourism to the mainstream, what are some of the lessons that can be adapted and adopted from some of the best approaches in Asia? The concept of responsible rural tourism focuses on tourism operations that are managed in a sustainable way, that it can continue to deliver the benefits for years to come for the community. Nonetheless, not all of these destinations have succeeded in developing their responsible rural tourism products. We can also learn from approaches that have not been successful.

According to Lane (1994), rural tourism is the most common form of tourism offered in rural areas and typically has the following characteristics – located in remote areas, small-scale enterprise, wide-open spaces, closely associated with nature, heritage, 'traditional' societies and practices. In addition, rural tourism destinations essentially have distinct characteristics – low levels of tourism development, and opportunities for visitors to directly experience the local economy which is mainly agricultural and/or focused on natural environments or non-urban settings (Frochot, 2005; Irshad, 2010). Rural tourism is rural in scale in terms of buildings and settlements, traditional in character, growing slowly and organically, connected with local families and representing the complex pattern of the rural environment, economy, history and location. These characteristics are evident in rural tourism destinations in many parts of Asia. Hence, the understanding of the development of a healthy and sustainable rural tourism destination is essential if Asia wishes to continue to attract and be a regional leader in responsible rural tourism.

The research of Jurowski (2008) has already indicated that the tourism industry's interest in appearing to be 'green' or 'sustainable' has increased in exponential proportions. Subsequently, the growth of the tourism industry through the years has created an increasing amount of stress economically, socially and environmentally as the carrying capacity or limit of these destinations is not checked or adhered to, as indicated in many studies over the past decade (Dlamini, 2013; Goodland, 1992; Hall, 2004; Sharpley, 2000; Vehbia & Doratlia, 2010).

Nonetheless, the concept of rural tourism has melded with mainstream tourism in many destinations in Asia. As a result, rural tourism as a niche tourism product has lost its distinctness in many parts of the world (United Nations Environment Programme [UNEP], 2010) including in Asia. Mass tourism in rural destinations will result in more long-term negative impact than the positive outcome it is expected to bring. Without careful attention to the balance between the volume and type of rural tourist activity and the sensitivities of the limit of the resources being developed, tourism projects can be not only environmentally harmful but also economically and socioculturally self-destructive (Mbaiwa, 2003). Understanding these complex dimensions that work in Asia, which may be similar or different from other parts of the world, is critical for the global sustainability of rural tourism destinations. Hence, rural tourism in Asia has its own uniqueness that attracts tourism from the Western world to experience the harmony between culture, environment and the micro-economy of the local community.

Conceptualizing Sustainable Tourism and Responsible Tourism

Sustainability is the ability to maintain at a certain rate or level. Sustainability avoids the depletion of natural resources in order to maintain an ecological balance, for example, in the pursuit of global environmental sustainability. As defined in the Brundtland Report (1987) entitled 'Our Common Future', sustainable development is, 'development that meets the needs of the present without compromising the ability of future generations to meet their own needs'. Most sustainability definitions embed economic development and social equity as important dimensions. Hence, sustainability is not just environmentalism. The Brundtland Commission successfully unified these three pillars of sustainability in the world's development agenda. Thus, this holistic approach in finding the right equilibrium is critical for lasting prosperity. In short, achieving sustainability does not mean that the destination's quality of life will be lost. What sustainability does is just change the mind-set and values towards a less consumptive lifestyle in any destination. This lifestyle is critical to the success of the many rural tourism destinations in Asia. All these changes will result in destinations embracing global interdependence, economic viability, social responsibility and environmental stewardship.

The triple bottom line focuses on the 'Three-Ps' of sustainable integration with equitable people (social) and the profit (economy); viable planet (environment) and the profit (economy); and bearable planet (environment) and the people (social). This is outlined in Figure 1.1. Nonetheless, in many destinations in Asia, a fourth 'P' has become another important dimension for sustainability – Politics. The political dimension can rock the equilibrium of sustainability. This dimension can include organization and governance, law and justice, communication and critique, representation and negotiation, security and accord, dialogue and reconciliation, and also ethics and accountability. These are all critical to the sustainability of any destination. Other similar concepts such as sustainable tourism, responsible tourism, rural tourism, niche tourism and slow tourism are all offshoots of this sustainability and sustainable development concept.

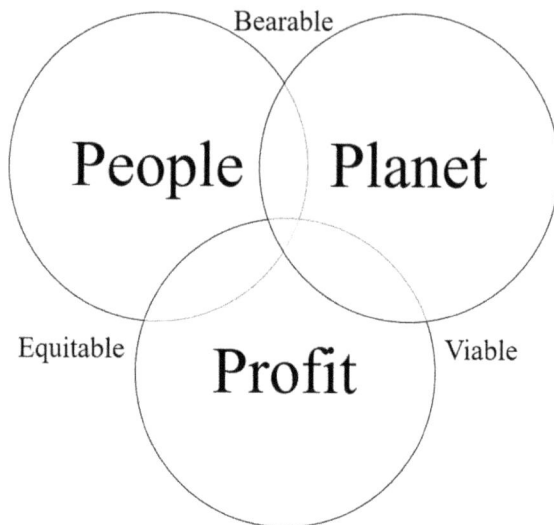

Figure 1.1 Triple-bottom line

The UNEP and United Nations World Tourism Organization (UNWTO) (2005: 11–12) defined sustainable tourism as, 'Tourism that takes full account of its current and future economic, social and environmental impacts, addressing the needs of visitors, the industry, the environment and host communities'. Hence, the concept of sustainable tourism makes optimal use of environmental resources that constitute a key element in tourism development while maintaining essential ecological processes and helping to conserve natural heritage and biodiversity; respects the sociocultural authenticity of host communities; conserves their built and living cultural heritage and traditional values; contributes to intercultural understanding and tolerance; and ensures viable, long-term economic operations by ensuring a fair distribution of socioeconomic benefits to all stakeholders, including stable employment and income-earning opportunities and social services to host communities, and contributing to poverty alleviation.

Then the following questions appear: 'If you are sustainable, does that mean that you are responsible? If you are responsible, does that mean that you are sustainable?' Sustainability is merely the goal. This goal can only be conceived and achieved by people taking responsibility together to achieve it. Thus, responsible tourism which originated from the concept of responsible development is a more operationalizable definition. The modern tourists of today have evolved in terms of their expectations when they visit a destination. As indicated by Nair and Azmi (2008), this new wave of tourists is saying 'no' to mass tourism, irresponsible operators and resorts that are destroying the local environment. These tourists want real quality experience. They want to know that the shower they are taking is not depriving a village of water. That the hotel they are staying at is not robbing the locals of their livelihood, or that their very presence is not offending the local communities. This is what responsible tourism is clearly about.

Responsible tourism was defined in Cape Town in 2002 (International Conference on Responsible Tourism in Destinations (ICRTD) (2002) alongside the World Summit on Sustainable Development. According to the definition, responsible tourism is: 'Tourism that provides quality travel experiences by maximizing the benefits and opportunities to local communities, minimizing negative social or environmental impacts, and helping local communities conserve fragile cultures and habitats or species'. Responsible tourism operations are managed in such a way that they preserve the local environment and culture so that it can continue to deliver the benefits for years to come. In short, 'responsible tourism is about using tourism to make better places for people to live in and better places for people to visit, in that order' (Goodwin, n.d.). This philosophy is evident in many successful responsible rural tourism destinations in Asia.

Hence, sustainable tourism is different from responsible tourism in that the latter focuses on what people, businesses and governments do to maximize the positive economic, social and environmental impacts of tourism. It is about identifying the important issues locally and addressing those and transparently reporting progress towards using tourism for sustainable development. Responsible tourism facilitates the process towards the bigger sustainable tourism agenda. Thus, all stakeholders – tourist, local community, operators, governments and non-governmental organizations (NGOs) – have to be responsible for achieving this agenda. Nonetheless, these two terminologies – sustainability and responsibility – are used interchangeably in the chapters despite the distinct differences.

According to Anderson (2011) and Nair (2018), in many rural destinations, the 'three-pillars' of sustainability quite often obscure the real relationship between the economic, the social and the environment. Can they then be regarded equal in all ecosystems? The environment is the physical reality all life depends on. The social is about one of the species (which includes the community) within the environment, our own, organizing itself; the economic is in turn one sub-set of the social. Hence, in rural destinations where the environment (the flora and fauna) is the primary product (or attraction) each of these dimensions should be nested within the next dimension – economic within social within environmental. Similarly, in destinations where the social (the local community) is the main pull factor for tourists to visit, environmental and economic dimensions will be within the subset of the social dimension. Thus, in all responsible rural tourism destinations, the question of which dimension

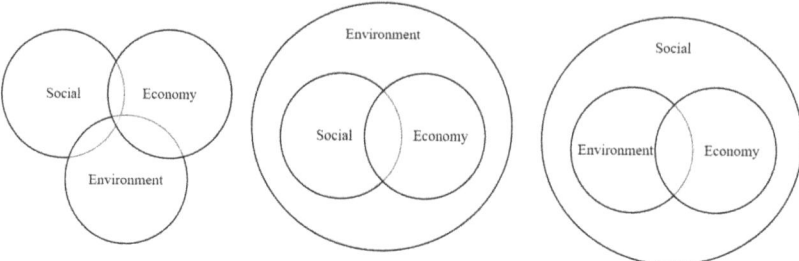

Figure 1.2 Knocking down the three pillars of sustainability

takes precedence will vary accordingly as can be seen in the case studies presented in the book (see Figure 1.2).

The first Venn diagram in Figure 1.2 is ideally conceivable for destinations where the three dimensions of sustainability have an almost equal importance. For diagrams two and three, the importance is put on the environmental and social dimensions, respectively. The case studies presented in this book have adopted the three different approaches in managing the dimensions of sustainability.

Defining Rurality and Rural Tourism

Defining rural tourism may seem obvious, as 'tourism that takes place in the countryside'. Nonetheless, this definition does not include the complexity of the activities and the different forms and meanings developed in different countries, as well as the number of protagonists participating in rural tourism as described by Bramwell and Lane (1994) over the past two decades. The meaning and context of rurality differs from one country to another, very much depending on the economic development phase of the nation. In many parts of the world, the word 'rural' has a negative connotation with which most local community do not want to be associated with (Nair et al., 2014). For these people, 'rural' basically means 'backward', 'unsophisticated' and 'rough'. The word 'rural', has either favourable or unfavourable connotations. In a derogatory sense, it may mean 'provincial', 'boorish' or 'crude'; and in a favourable sense, it may suggest gentry, ruggedness or a homelike rural charm, that is, rustic simplicity. Hence, 'rural tourism' may need to evolve to a term that is more favourable.

As to whether the term 'rural' is baggage or a sign of a privileged lifestyle depends for instance on the state of the rural economy. Furthermore, it depends on the standpoint; a term like 'rural economy' can simultaneously mean the same and different things to a farmer in the Caribbean than to his colleague in Southeast Asia. However, the term 'rural' is probably more often used to mean not urban, which in most parts of the world is rather the exception and used as a marketing tool for destinations that are rural. That is where time stood still – an ideal that has so much tourism appeal; but the flip side of the coin is that where time stood still, poverty may have crept in as the means of production yielded little for the people. This was further asserted by Lane (1994: 17) that, 'Traditional, agriculture and forestry were central to rural life. They were the major employers of labour, the main sources of income within the rural economy and indirectly had a powerful influence on traditions, power structures and life styles'.

As in many other regions, developing tourism in rural areas faces major obstacles in Asia. There is huge gap between the requirements of typical tourism products and the characteristics of rural areas. This was outlined by Holland *et al.* (2003) and shown in Table 1.1. Hence, more efforts are required for the rural areas to be able to meet the demands of tourism development. Rural destinations require primary attractive tourism products that will attract the visitors to brave the hardship to reach the destination, supported by good accessibility and availability of skills. In addition, the institutional and political problems in the country can assume great importance in the success of rural tourism destinations (remember the fourth 'P'). Government support is critical in terms of investment, appropriate regulation and marketing, as well as including rural tourism as an important economic strategy.

Thus, while the term rural tourism is used without hesitation in the context of many developed nations in Europe, it is obvious that it may have different connotations in other parts of the world, for instance where

Table 1.1 Determining the gaps between tourism development and characteristics of rural areas

Common requirements for tourism development	Common characteristics of rural areas
A product, or potential product	Variable. May have a high-value unique selling point, may be an attractive desired location for travellers from cities, may have little to offer.
Access – transport infrastructure, limited distance, limited discomfort	Distant from cities, poor roads, few trains/buses/planes
Investment in facilities	Limited access to financial capital, affordable credit and private investment.
Skills in service, hospitality	Low skills (skills migrate)
Regular and quality inputs, e.g. of food and other supplies	Undeveloped commercial production, distant from markets
Marketing skills	Distant from marketing networks
Clustering of tourism products to create a 'package' holiday	Lower concentration of tourism products in one place
Government investment	Low priority for governments, particularly tourism/trade ministries.

the economic relations of host and guest are very unequal especially in many parts of Asia, as can be seen in the subsequent chapters in this book. In most parts of Asia, the focus of the government is to somehow 'develop' the remote rural through community-based tourism focusing heavily on traditional culture, arts, crafts and traditional ways of life. The intention is therefore to appeal to an international market and also reintroduce rurality to the millennial youth domestically.

While it is commonplace to mark that the term 'rural' is opposed to 'urban', the geographical term 'rurban' is also used to denote the merging of the bucolic and other characteristics that are found in rural areas as opposed to the concrete jungles found in urban areas (Kumar & Sikarwar, 2017). In many developing countries in Asia, the rural landscape has evolved. Therefore, the term rural, though denoting and connoting different things, has to be perforce used, but can be differentiated from village tourism or countryside tourism.

Subsequently, the concept of rural tourism has emerged to encompass many forms of tourism. Rural tourism may include nature-based tourism, ecotourism, community-based tourism, agro-tourism, and many more. According to Rawat (2015), rural areas can be defined based on the various circumstances considering its applicability within a study context. However, there is no standard outlined definition for rural areas. The definition frequently depends upon the view point of the researcher, study context and the geographical periphery (see Table 1.2). These studies normally highlights the definition of rural areas based on the indicators of rural tourism as reflected in the masterplans of the selected countries. Furthermore, Simkova (2007) contends that defining rural areas collectively is a much-discussed issue because countries and organizations have different indicators to define rural areas.

As Table 1.2 demonstrates, definitions of rural areas are considerably different from each other: some are more or less based on density; some are associated with distance; and some are highlighted based on the influence of the service industry which in a way brings urbanization, and thus, no longer considered as rural areas.

The lack of consensus on the definition for the concept of rural tourism is evidenced by the number of different definitions used in the literature either by international- or regional-level organizations such as the Organization for Economic Co-operation and Development (OECD), World Tourism Organization (WTO) and the European Union (EU) or by individual authors/researchers.

The OECD (1994: 34), defines rural tourism

> as being located in rural areas, and as being functionally rural... that is, firmly based on the rural world's special features of open space, contact with nature, rural heritage and society. Its scale should be in keeping with the landscape and settlements in which it operates: those settlements are normally of fewer than 10,000 people. While including farm tourism

Table 1.2 Definition of rural areas as indicated in rural tourism masterplan of the country

Countries	Definitions
Australia (1994)	Areas where population is less than 1000 living exclusively on the service industry such as holiday resorts.
Austria (1994)	Cities with less than 5000 people.
Canada (1994)	Areas where the population is less than 1000 with a density of less than 400/square kilometres.
Denmark (2007)	Small towns with up to 1000 inhabitants.
France (1994)	Cities where population is less than 2000 living in a close area with a minimum distance of 200 m from each other.
India (1994)	Areas where the population density is less than 10,000.
Thailand (2014)	An area with the remaining population after calculating the urban population.
Malaysia (1994)	Areas where the population is less than 10,000.
Indonesia (2014)	An area with the remaining population after calculating the urban population.

within its remit, its overall focus should also encompass the whole range of suitable businesses and settlement types in the countryside. Its aim should be to help ensure the long-term sustainability of the life of the region: it should be a force for the conservation of rurality rather than a force for urbanisation.

Further, the UNWTO (Aref & Gill, 2009) defines rural tourism as a tourism product 'that gives to visitors a personalized contact, a taste of physical and human environment of countryside and as far possible, allows them to participate in the activities, tradition and lifestyles of local people', whereas the EU (Eurostat, 2005, cited in Lebe & Milfelner, 2006) defines rural tourism as 'the activities of persons travelling to and staying in rural areas/without mass tourism/other than those of their usual environment for less than one consecutive year for leisure, business and other purposes (excluding the exercise of an activity remunerated from within the place visited)'.

Nair *et al.* (2014) redefined rural tourism definition for Malaysia by analysing the multiple dimensions and complexities of the definitions in other developing and developed nations. The five core dimensions that emerged from the content analysis of rural tourism definitions in the developed and developing economies are 'location characteristics', 'purpose of visit', 'attractions/activities', 'sustainability' and 'stakeholders'.

Over the past decades, as part of rural tourism offerings in Asia, community-based tourism (CBT) is seen as an alternative form of community development strategy in the context of rural areas. It is considered as an appropriate strategy to tackle the rural needs which deal with the grounded issues of locals and their fundamental requirements. In this setting, the majority of locals are engaged in to compensate the rural socioeconomic issues, by providing them an access of right over their resources and power on the decision-making process delivered by the service industry (Suansri, 2003). The implementation of CBT is sustainable when its primary benefits are witnessed by and for the locals, and later, to the various stakeholders, working under a collaborative environment (Jamal & Stronza, 2009; Pearce, 1992; Reed, 1997; Wyllie, 1998). As further indicated by Simpson (2009), CBT seeks to create local enterprises that provide livelihood benefits to communities while protecting indigenous cultures and environments, thus making these enterprises key projects for sustainable tourism. The ideal CBT approach ensures that a significant proportion of the benefits must be received and led by local residing in that area to protect their well-being, has equitable participation, and there is sustainable use of natural resources (Iorio & Corsale, 2014; Stone, 2015)

In some destinations in Asia, CBT tended to involve tours and overnight stays, whereas 'community-involved tourism' (CIT) could be regarded as CBT plus the community producing items for the tourism supply chain such as food, furnishings and souvenirs (Gillbanks, 2017). Therefore, CBT or CIT are considered as one of the major rural tourism products as can be seen in many of the case studies presented in the subsequent chapters. CBT is mainly (not always) operated under rural settings in the context of Southeast Asia and Asia. Suansri (2003) further added that CBT is 'practised and owned by the community, for the community, to improve their social and economic aspects of life'.

Although most rural tourism development generally begins with a governmental policy or support programme, many successful rural tourism destinations in Asia emerged from the initiatives and efforts of the local community themselves, through an association of interests. All rural tourism development ideally must be directed by a responsible tourism masterplan. All stakeholders involved in managing rural tourism must be provided with professional training to ensure commitment to quality. Up-to-date statistics on rural tourism are needed to measure its scale and impacts on agriculture and rural life. As can be seen in many of the case studies in Asia, rural household economies use different strategies such as diversification, migration and the intensification of production to survive and manage their risks (Ellis, 2000). Many studies (Alobo Loison, 2015; Berhaus et al., 2007; Mushongah & Scoones, 2012) have indicated that diversification is recognized as a core strategy of rural livelihoods that is critical for the success of rural tourism. Through diversification, households can engage in a variety of activities to survive and enhance their well-being (Ellis, 2000).

In short, preparing for tourism requires that a rural community in Asia take a critical look at itself. The whole new concept of 'overtourism' (Goodwin, 2017; Milano *et al.*, 2018) and 'undertourism' (Beabout, 2019; Weissmann, 2019) can impact the development or underdevelopment of rural destinations for tourism. Rurality indeed can transform rural tourism as the modern tourists of today look for this harmonious authentic experience. Authentic rural tourism attracts visitors who like to see things just the way they are – not staged. It is about getting visitors onto the back roads looking for heritage structures, unique gastronomy, traditional crafts and music, etc. More importantly responsible rural tourism in Asia should portray the 'Asianess' of the tourism offerings, based on its distinct imageries. The authors have been asked to ponder whether there is such thing as a quintessential rural tourism in Asia based on its traditional philosophy and relationship between humans and nature. It is contended that rurality in Asia is predominantly but not merely confined to picturesque and quaint villages in the countryside or formal protected areas. Instead, it is further contended that rurality in Asia is the embodiment of human adaptation to nature that has created socio-ecological processes that put spiritual values at the core of its worldview and cosmology. Whether or not Asian rurality has been able to shape an essential form of responsible rural tourism, albeit being challenged by globalization and modernity, is explored by the authors in the subsequent chapters.

Structure of the Book and Chapter Synopsis

This book, consisting of the introduction, the 12 chapters, and conclusion, reveals three main themes. Rather than present the synopsis of the chapters in a linear pattern, they are introduced based on three distinctive themes.

Theme 1: Harmony Between Humans and Nature in Moulding Rural Tourism in Asia

The first theme is how the Asian development philosophy has shaped 'rurality', and in turn, the form of rural tourism in Asia. From a country-wide perspective, the chapter by Schiffling *et al.* (Chapter 2) recognizes the parallel between Asia's development philosophy and the principles of responsible tourism. Based on the pillars of gross national happiness (GNH), the authors present Bhutan's alternative development philosophy, which focuses on the protection of the rural environment and resources, the preservation of culture and the spiritual wellbeing of the community instead of the mainly economic development agenda associated with the gross domestic product (GDP) that the rest of the world uses. Central to Bhutan's GNH is the Asian philosophy of harmony between humans and nature, the guiding principles provided by the pillars of the GNH are shown by the authors to be consistent with the Cape Town Declaration on Responsible Tourism.

The resulting low-impact, high-value tourism, as argued by the authors, has been delivered by the country's top-down institutional structure and moulded by the adherence to the GNH pillars. By deliberately being faithful to an alternative (and Asian) development path, the government's focus on responsible tourism has been shown to cascade to the mainly rural tourism destinations that are popular with international tourists. However, the authors also highlighted that *realpolitik* might pose a threat to this equilibrium given that visitors from neighbouring India are unrestricted. A parallel to this scenario is the massive 'exodus' of tourists from the densely populated Indian city of Kolkata (population 14.4 million) to the ancient Himalayan kingdom of Sikkim during weekends. As overtourism is becoming a major concern in Asia, especially with the growth of intra-Asian tourism driven by China and India, how the governance and management of tourism in Bhutan respond to this looming threat, which is not only confined to the destination level but is likely to be detrimental to the country's unique development philosophy, will provide valuable lessons related to responsible rural tourism.

At the destination level, Sofield and Li provide a rich commentary on the merits of the Asian philosophy of harmony between humans and nature in their chapter on cave tourism in Guilin, China (Chapter 3), The authors contend that the notion of responsible rural tourism should not be solely viewed from the Western biocentric philosophy of separating humans from nature, in order to protect the natural, pristine and wilderness. Alternatively, the authors offer a Chinese philosophical standpoint to justify that the current use of technology to facilitate visitor access and movement is an extension of the artificial physical modifications that have been added for centuries to the caves in Guilin and other parts of China, in the form of calligraphy (poems) being carved on cave walls and the erection of temples and statues at sacred spots. From the Western perspective, such artificial modifications would be interpreted as a form of 'Disneyfication'. However, Sofield and Li challenge this view by arguing that the artificial elements of cave tourism in Guilin to create entertainment and fun for the mainly domestic tourists are not a reflection of human dominance but rather of co-existence with nature. Hence the radical assertion offered by Sofield and Li has contributed to a blurring of what constitutes responsible rural tourism and its universal interpretation. From the managerial perspective, however, it is posited that the contemporary interpretation of the Daoist philosophy might pose serious challenges in efforts to reconcile the impact from the influx of domestic tourists to fragile rural tourism destinations and their social construction of nature.

Theme 2: An Essentially Asian Form of Rural Tourism

The second theme is the quest for an essentially Asian form of rural tourism. Having established the difference in the development philosophy of Asia, it is then pertinent to ask whether the development of rural tourism is grounded on this philosophy, and if so, has it shaped the creation of an essential Asian form of rural tourism. Towards this end the case study on tea tourism in Japan by Jolliffe and Nakashima (Chapter 4) provides a compelling analysis of how tourism that is deeply rooted in the Japanese mindset could provide a viable solution to revitalize rural areas that have suffered extreme depopulation. In Japan the massive outmigration of the younger population to the cities has become a national problem, leaving mainly the elderly living precarious lives in the rural areas due to the breakdown in the community structure and social cohesion (Walia, 2019). In this context, the authors describe how tea tourism based on the intricacies of the Japanese tea ceremony has introduced a low-impact, high-value tourism in response to the needs of the urban population's quest for *Furusato* (longing for 'old' village). In essence, the sacred tea ceremony in the tea villages is an expression of the serenity of the rural Japanese setting that had been moulded by its socio-ecological production landscape to create the celebrated *Satoyama* reality and also tourist imagery.

Based on their case studies, the authors also recommend future research into the demand side of tea tourism which would be able to shed more light into the potential of tea tourism in luring back the youth to operate tourism-related businesses along the supply chain. Significantly this chapter has also shown how the development of an integrated form of rural tourism, based on the uniqueness of its sense of place and rural imagery, has the potential to shape responsible rural tourism in Asia. However, the threat to the 'scale' of tea tourism, as an essential characteristic of rural tourism (Lane, 1994), is expected to be real in the near future as Japan continues with its recently open approach to international tourism. With the sudden influx of mass tourists especially from China, and their inevitable search for the 'back stage' (Hall *et al.*, 2003), it is envisaged that the possible threat of 'overtourism' is likely to have significant management implications on the sustainability of tea tourism, given the precariousness of rural life in a country that has the highest ageing population rate in the world.

Sense of place is an essential element in rural tourism as it provides the intrinsic character of the place that shapes the meaning that residents and visitors put to the place (Ngo & Brklacich, 2014). Thu's chapter (Chapter 5) on the interface between garden tourism and monsoon tourism in Hoi

An, Vietnam provides a captivating narrative of how Asia's natural phenomenon, i.e. the tropical monsoon that is associated with heavy and prolonged rain, and even flooding, has created a unique sense of place to be promoted to tourists. Through the eyes of Asian cinema (Pugsley, 2016), rain is considered a powerful as well as an empowering force. As such, rain often provides the setting and mood for romantic scenes in Bollywood movies while the battle scenes in the epic films by the late and renowned Japanese film director, Akira Kurosawa (*Rashomon, Ran, Seven Samurai*) were all depicted against the backdrop of a dramatic torrential downpour. As a form of tourism business, monsoon tourism has been deliberately promoted by Asian countries such as Bangladesh and Malaysia, with varying degrees of success, to overcome the seasonality problem imposed by such natural phenomenon. The findings of Thu's case study show that an attractive sense of place has been created by the organic farms and gardens surrounding Hoi An World Heritage Site, with the monsoon season providing a dramatic and romantic backdrop. More importantly, Thu captures the nuances of the demand side of monsoon tourism in tandem with farm/garden tourism that is unmistakably associated with the romantic gaze (Urry & Larsen, 2011).

Viewed from a wider perspective, the growing popularity of monsoon tourism offers an intriguing prospect of selling Asia's natural phenomenon as a distinct rural tourism offering in tandem with disaster tourism (Tucker *et al.*, 2016). Interestingly, it has been highlighted that the population of Asia is 25 times more exposed to the risk of natural disasters than the population of North America and Europe (ADB, 2013). In 2018, natural disasters such as tsunami, earthquakes, landslides and floods resulted in extensive damage to properties and the loss of human lives including tourists in many parts of Asia (especially the earthquake in Lombok and tsunami in Sulawesi, Indonesia). Without offering an ethical or moral judgment, it can be surmised that Asia's natural disasters have provided the impetus for the rise in disaster tourism in the region.

Nonetheless it is suggested that instead of focusing on the catastrophe and its aftermath as the tourist gaze, a more responsible form of tourism could be created by emphasizing Asia's adaptive approach to co-existing with nature that has given birth to the term 'nature-based solutions' in confronting natural disasters (ADB, 2013). In the face of climate change, adaptation and mitigation approaches using natural solutions are increasingly regarded as being more effective solutions. In the Hoi An case study, the use of sustainable agriculture principles in the creation of organic farms could be viewed as the local-level actions that support nature-based solutions. On a wider scale Japan is at the forefront of showcasing disaster risk reduction (DRR) approaches and models including the creation of the Sanriku Fukko National Park that was specifically designed with nature trails that could be used to evacuate residents and visitors in the event of a tsunami (ADB,

2013). DRR-related tourism has the potential of cementing Asia's contemporary approach in ensuring harmony between humans and nature, and providing a model for responsible rural tourism in an increasingly uncertain world brought about by climate change. Against this backdrop, the monsoon tourism experience in Vietnam can be regarded as an essentially Asian form of responsible rural tourism that showcases human adaptation to nature and the humble acceptance of its powerful forces.

As demonstrated by the tea villages in Japan (Chapter 4) and gardens/organic farms in Vietnam (Chapter 5), the tourist imagery is closely related to their natural and cultural resources. In the chapter on responsible rural tourism in Kerala (Chapter 6), George *et al.* demonstrate that a systematic tourism planning institutional framework was able to optimize the tourism supply chain in strengthening the linkages between the tourism industry and local producers. The authors reveal that, given that Kerala is an established spice capital, this relationship has been critical in adding value to the tourism industry through wellness tourism based on the ancient Indian herbal massage and traditional treatment called Ayurveda. Central to this effort is the establishment of a formal tourism administration structure that cascades responsible tourism principles and practice from the provincial (State) level to the destination level. This top-down institutional structure has been instrumental in empowering the local community and consolidating the destination's unique selling proposition (USP). A clear division of power and responsibility has minimized the problem of overlapping jurisdictions besides allowing the State Level Responsible Tourism Committee (SLRTC) to provide a clear policy direction that is complemented by the role of a tourism training institute mandated to provide training and capacity building.

The authors also highlighted that appropriate to the needs at the destination level, the Destination Level Responsible Tourism Committee (DLRTC) provides 'handholding' for the local operators and local producers and community. By analysing the tourism supply chain at the destination level, the DLRTC was able to identify the economic leakages and subsequently nurture a strong business collaboration between the accommodation providers/eateries with the local producers and farmers. In the wider context, the wellness tourism element of Kerala has provided a considerable value add to one of the most popular tourist destinations in Asia, in which the unique houseboats, cruising the backwaters facing the Arabian Sea, are central to the tourist image. The main lesson from the Kerala case study is how an established rural tourism destination could be further upscaled by leveraging on the synergistic relationship between the local herbal/spice industry and wellness tourism, to penetrate the lucrative global wellness tourism industry that is worth around USD439 billion annually (compared to the USD50–USD 60billion for MICE) (SRI International, 2016), without compromising the principles of responsible rural tourism.

Theme 3: Tourism That Takes Place in Rural Areas

Having presented an intrinsically Asian form of responsible rural tourism that is rooted in the philosophy of co-existence between humans and nature, it can be seen that its characteristics have not deviated conceptually from Lane's (1994) constructs. With the exception of the increasingly dysfunctional rather than traditional rural areas, intrinsically Asian rural tourism is still on a small scale in nature and is pastoral in function, besides exuding the perception of the rural idyllic within an Asian socio-ecological production landscape.

Nevertheless, and in response to the ever-increasing demand for recreation from the dense population living in its cities as well as businesses, opportunism has seen the development of a wide spectrum of attractions in the rural and peri-urban areas of Asia. 'Family fun' is the vogue in Asia as its growing middle class is increasingly searching for entertaining and even challenging tourism experiences beyond just recreation, thus giving rise to the development of theme farms, adventure parks, agritourism trails and even 'mimicking nature' theme parks, which makes it much easier to define 'what is not rural tourism rather than what it is' (Sorenson & Nillson, 2003). In this light, the third theme of this book is tourism that takes place in rural areas and whether it is a 'deliberate' form of responsible rural tourism. Building on Weaver's (1991) dichotomy of 'deliberate' versus 'circumstantial' alternative tourism, Sorenson and Nillson (2003) argue that rural tourism is one of the many variants of alternative tourism. By the same token, it is contended that the characteristics of 'deliberate' alternative tourism (as well as 'circumstantial' alternative tourism) could be applied in the context of rural tourism. Central to the concept of 'deliberate' responsible rural tourism is that it should be *planned and delivered in this way*.

Agritourism is regarded as one of the subsets of rural tourism and the case study of agritourism in the Philippines discussed by Mena and Sy (Chapter 7) offers a fascinating insight into the way tourism is taking place and spreading to the rural areas of Asia. From the lucid discussion given by the authors, it could be surmised that agritourism has been institutionalized in the Philippines through the setting up of the Philippines Farm Tourism Industry Development Coordinating Council (PFTIDCC), which is supported by around a dozen Federal agencies. Furthermore, an accreditation system has been developed that has led to the accreditation of more than 30 agritourism sites since 2012. Despite this, the authors argue that transforming rural resources into rural tourism products remains a challenging undertaking given the lack of local capacity in respect to the core skills required of tourism. The day farms and farm resorts in the Philippines fit the description of 'family fun' attractions and activities that are also mushrooming around the primary and secondary cities in Asia. Notwithstanding the commonality in terms of concept and spatial development, the creation of 'family fun' agritourism attractions in the rural and peri-urban areas of the Philippines is unique because they are coordinated and regulated by the government as a concerted attempt to add value to the agriculture sector through rural tourism. As agritourism becomes institutionalized, the authors reveal that the multiple agency approach has been able to add depth to the tourist experience, carry out capacity building in a systematic manner and ensure quality assurance through the accreditation system. On the other hand, it is likely that similarity and the lack of differentiation might occur as a result of the top-down and regimented approach. Nonetheless agritourism and the clustering of its variety of products to offer a 'family fun' experience is expected to dominate the future of rural tourism in Asia, which will provide not only a fertile area for research but also innovative solutions for the revitalization of rural areas in the region. Despite this, the question remains whether there is a 'deliberate' attempt in introducing a responsible form of rural tourism beyond improving the livelihood of the agritourism operators. This apprehension stems from the authors' finding that the agritourism operators have yet to fully understand the demands of running a tourism business let alone the principles of responsible tourism that they have to adhere to.

In the same vein, the CBT project in Coruh Valley, Turkey which Karacaoglu and Yolal present in Chapter 8 is a top-down pro-poor tourism project designed and driven by outsiders. Conceived by the United Nations Development Programme (UNDP) and the government, the delivery was led by an industry partner to provide a 'quick win' solution to rural poverty in the region over a four-year project span. The authors' field interviews reveal both the success in using tourism as catalyst for revitalizing the economically depressed Coruh Valley in the short term as well as the apprehensions of sections of the community in terms of the long-term viability of the project. Among the local community's negative perceptions were the potential encroachment by outside investors, constant threat from the construction of hydroelectric dams and the limited coverage of the project. Notwithstanding the introduction of a training programme that had empowered the local women and youths, the authors identified that the lack of a more comprehensive capacity building programme had impeded the setting up of a community-based governance structure that could wean off their overdependence on outside help. Both the agritourism project in the Philippines and the private sector driven CBT in Turkey reveal concerns over the lack of 'buy in' from the local communities that might affect their understanding of the form of rural tourism offering that they are 'expected to deliver' and their long-term commitment to the project. In essence, both of these top-down approaches could be regarded as 'circumstantial' responsible rural tourism.

To ensure a long-term commitment to responsible rural tourism, social cohesion – specifically community cohesion – is a critical success factor. Towards this end, CBT provides the vehicle for rural communities to participate actively in tourism in which community involvement, community control and community benefit are central to its success (Mitchell & Ashley, 2009; Scheyvens, 2002; Tosun, 2000). Ngo and Nguyen's chapter on CBT in Vietnam (Chapter 9) argues that community cohesion is critical to the success of the two CBT case studies expressed through ethnical homogeneity and kinship. Fundamentally, ethnical homogeneity and kinship represent non-Western values of 'good governance' (Graham *et al.*, 2013) that the authors in this chapter regard as being equally as important as the CBT cooperative (Western value) in the governance structure of the CBT projects. As such the authors argue that showcasing the local community's cultural practices and the ensuing cultural immersion as an essential part of the CBT's experience has contributed to the preservation of the community's values and cultural heritage, which are less evident in the case studies from the Philippines (Chapter 7) and Turkey (Chapter 8).

Building on the issue of capacity building in responsible rural tourism, Bouchon's analysis (Chapter 12) of a donor-initiated, private-sector-driven tourism project in Timor Leste reveals a common thread in the way capacity building is often limited to skills advancement. More importantly the author highlights the complexity of power play in rural tourism, especially the fact that power usually resides not with the community but mainly in the hands of outsiders such as donors, investors and politicians. The Timor Leste commentary also supports the recurring theme that traditional power structures are often bypassed in the decision-making process and in the governance structure of rural tourism, hence limiting the development trajectory towards a responsible form of tourism. Furthermore, Bouchon also identifies the need for a systematic tourism planning process that encompasses a multi-stakeholder collaboration as the required foundation towards 'deliberate' responsible rural tourism.

Good governance and what constitutes good governance is a critical component of responsible rural tourism as elucidated in many of the chapters. Also, it should be noted that there is a difference between government and governance (Goodwin, 1998). While there are universal principles of good governance that include transparency, accountability, empowerment and rule and law and so forth, there are significant differences in Western values and non-Western values that influence their varying interpretation. For instance, Western values emphasize the rights of individuals but non-Western values regard communal obligation as being more fundamental. Likewise rule of law as a Western value contrasts with tradition and kinship in the decision-making process (Graham *et al.*, 2013; Shields *et al.*, 2016). Such contestations between Western and non-Western values are well captured in Goh and Nair's analysis of the governance framework of Belum-Temengor Forest Reserve, Malaysia (Chapter 13). As indigenous communities are among the key stakeholders of many rural tourism destinations, the formal and essentially top-down institutional framework, which is often laden with Western values, fails to incorporate their worldview and customary organization. Goh and Nair highlight that the lack of an effective channel of communication has suppressed the voice of the indigenous *Orang Asli* community in the decision-making process. As a consequence, it is not surprising to add that the *Orang Asli* in their enclaves at the Belum-Temengor Forest Reserve are still being degradingly 'exhibited' by tour operators. The main lesson from the Belum-Temengor case study is that having a formal governance structure that does not incorporate local values and rights will impede the development of 'deliberate' responsible rural tourism.

In a nuanced discussion of the 'buy in' process involving the local community, Hamzah's chapter on the critical success of CBT in Sabah, Malaysia (Chapter 10) traces the success of the project to the role of a succession of local champions within a close-knit community. Similar to the CBT case studies in Vietnam (Chapter 9), the author illustrates that ethnical homogeneity and kinship had formed the backbone of the informal governance structure of the Miso Walai Homestay in Sabah during its formative years. Notwithstanding the transformation of the CBT project's business model to include strong partnerships with related government agencies and specialist tour operators, Hamzah reveals that the management of the CBT project currently comprises the pioneer local champions and their offspring who have returned to the village after completing their tertiary education. It is evident from the author's longitudinal study that the succession of local champions had been instrumental in sustaining the common vision of developing low-density, low-impact tourism in synergy with active conservation in the form of community reforestation programmes.

Moreover, it lends weight to the argument that Asian values of good governance that recognize traditional and customary community decision-making processes are essential to the management of CBT projects to ensure local control and an equitable distribution of income (Goodwin, 1998). More importantly the findings of the study reveal that the CBT project had embarked on a 'deliberate' form of responsible tourism since its inception in the mid-1990s. Although the CBT project at Miso Walai Homestay was initiated by an international donor, it went through a process of bottom-up planning that involved the whole community along a tourism planning process that was systematic and inclusive. By not being dependent on the government, Miso Walai Homestay's endogenous governance structure has facilitated its delivery of a multi-dimensional form of responsible rural tourism from the perspective of the tourism industry, tourists and local community.

Although international donors and NGOs have leveraged CBT to (re) create sustainable livelihoods in relation to conservation projects, their project-based nature within a specific time span has been criticized as giving more emphasis on the conservation agenda while relegating the economic wellbeing of the local community (Butcher, 2007). In addition, the overdependence on technical advisors appointed by donors have persisted even for successful CBT projects (Harrison & Schipani, 2007). Despite these shortcomings, Sofield's insider perspective (Chapter 11) of the accomplishment of the Chambok CBT in Cambodia mirrors the critical success factors that shaped Miso Walai Homestay's adherence to responsible rural tourism principles. Central to the success of Chambok CBT is the transformation of the British NGO that initiated the project into a localized and independent NGO, MlupBaitong, the same way that KOPEL (Koperasi Pelancongan Mukim Batu Puteh or Tourism Cooperative) in Sabah had become independent of WWF (World Wide Fund For Nature) to chart its own common vision and development path. Correspondingly, the CBT management framework of both projects integrated the traditional power structure with a contemporary business model centred on the CBT cooperative to operate an endogenous governance structure, which in turn, is central to the delivery of a rural tourism product that encompasses the whole spectrum of responsible rural tourism.

Conclusion

Responsible rural tourism is indeed becoming an important agenda of the tourism industry in Asia. Rural tourism has contributed significantly in terms of income generation to developing countries in Asia given its ability to eradicate poverty and improve the quality of life of communities away from urban development. From the case studies in the subsequent chapters, it is obvious that the demand for the rural tourism product has significantly increased in recent years. Cooperation and collaboration among the government organizations, local community and the private sector is critical to sustain rural tourism. Thus, rural tourism can be used by the government as a tool and strategy for rural regeneration and diversification of under-developed areas. Central to the differentiation and sustainability of rural tourism in Asia is the creation of an essentially Asian form of tourist imagery and tourist gaze. Towards this end it is contended that the spiritual values of Asian rurality need to be preserved and reinvented in the face of the onslaught of tourists to rural areas in Asia. This is crucial, as the increase in the tourist arrivals may lead to merely more general tourism activities that is taking shape in rural areas rather than trying to focus on offering an essentially Asian form or responsible rural tourism.

References

Asian Development Bank (ADB) (2013) *The Rise of Natural Disasters in Asia and the Pacific: Learning from ADB's Experience.* Mandaluyong City, Philippines: Asian Development Bank Publication.

Alobo Loison, S. (2015) Rural livelihood diversification in Sub-Saharan Africa: A literature review. *The Journal of Development Studies* 51 (9), 1125–1138.

Anderson, V. (2011) Let's knock down the three pillars of sustainable development. See https://www.socialeurope.eu/lets-knock-down-the-three-pillars-of-sustainable-development (accessed 3 March 2018).

Aref, F. and Gill, S.S. (2009) Rural tourism development through rural cooperatives. *Nature and Science* 7 (10), 68–73.

Beabout, L. (2019) You've heard of overtourism. Here's why undertourism is just as important. *Green Suitcase Travel* March 2, See https://greensuitcasetravel.com/2019/03/undertourism (accessed 25 January 2019).

Berhaus, W., Colman, D. and Fayissa, B. (2007) Diversification and livelihood sustainability in a semi-arid environment: A case study from southern Ethiopia. *Journal of Development Studies* 43 (5), 871–889.

Bramwell, B. and Lane, B. (1994) *Rural Tourism and Sustainable Rural Development*. Clevedon: Channel View Publications.

Brundtland, G. (1987) Report of the World Commission on Environment and Development: Our Common Future. See http://www.un-documents.net/our-common-future.pdf (accessed 26 July 2019).

Butcher, J. (2007) *Ecotourism, NGOs and Development*. London. Routledge.

Dlamini, W. (2013) *Biological Diversity – The Spice of Life*. Swaziland National Trust Commission. See http://www.sntc.org.sz (accessed March 2013).

Ellis, F. (2000) The determinants of rural livelihood diversification developing countries. *Journal of Agricultural Economics* 51 (2), 289–302.

Frochot, I. (2005) A benefit segmentation of tourists in rural areas: A Scottish perspective. *Tourism Management* 26 (3), 335–46.

Gillbanks, D. (2017) Ownership, inclusion key to community-based tourism. See https://goodtourismblog.com/2017/06/ownership-inclusion-key-community-based-tourism/ (accessed 6 September 2019).

Goodland, R. (1992) The case that the world has reached limits: More precisely that current throughput growth in the global economy cannot be sustained. *Population and Environment* 13 (3), 167–182.

Goodwin, H. (2017) The challenge of overtourism. *Responsible Tourism Partnership Working Paper 4*. October 2017.

Goodwin, H. (n.d.) Responsible tourism. https://www.haroldgoodwin.info/responsible-tourism

Goodwin, M. (1998) The governance of rural areas: Some emerging research issues and agendas. *Journal of Rural Studies* 14 (1), 5–12.

Graham, J., Haidt, J., Koleva, S., Motyl, M., Iyer, R., Wojcik, S.P. and Ditto, P.H. (2013) Moral foundations theory: The pragmatic validity of moral pluralism. *Advances in Experimental Social Psychology* 47, 55–130.

Hall, C.M. (2004) Space-time accessibility and the talc: The role of geographies of spatial interaction and mobility in contributing to an improved understanding of tourism. In R.W. Butler (ed.) *The TALC (Vol. 2): Conceptual and Theoretical Issues*. Clevedon: Channel View Publications.

Hall, C.M. and Page, S.J. (eds) (2017) *The Routledge Handbook of Tourism in Asia*. London: Routledge.

Harrison, D. and Schipani, S. (2007) Lao tourism and poverty alleviation: Community-based tourism and the private sector. *Current Issues in Tourism* 10 (2), 194–230. DOI: 10.2167/cit310.0

Holland, J., Burian, M. and Dixey, L. (2003) Tourism in Poor Rural Areas. *PPT Working Paper*. No. 12.

International Conference on Responsible Tourism in Destinations (ICRTD) (2002) The Cape Town Declaration, Cape Town.

Iorio, M. and Corsale, A. (2014) Community-based tourism and networking: Vicri, Romania. *Journal of Sustainable Tourism* 22(2), 234–255.

Irshad, H. (2010) *Rural tourism: An Overview*. Alberta, Canada: Government of Alberta Agriculture and Rural Development Publication.

Jamal, T.B. and Stronza, A. (2009) Collaboration theory and tourism practice in protected areas: Stakeholders, structuring and sustainability. *Journal of Sustainable Tourism* 77 (2), 169–189.

Jurowski, C. (2008) A tool for improving the sustainability of tourism industries. In *Proceedings of BEST EN Think Tank VIII*. Izmir, Turkey, June 24–27, 2008.

Kumar, V. and Sikarwar, S. (2017) Smart concepts for integrated rurban development of historical towns in India: Case of *Panipat, Haryana*. In F. Seta, J. Sen, A. Biswas and A. Khare (eds) *From Poverty, Inequality to Smart City*. Springer Transactions in Civil and Environmental Engineering: Singapore

Lane, B. (1994) What is rural tourism? *Journal Sustainable Tourism* 2 (1–2), 7–21.

Lebe, S.S. and Milfelner, B. (2006) Innovative organisation approach to sustainable tourism development in rural areas. *Kybernetes* 35 (7/8), 1136–1146. doi:10.1108/03684920610675139.

Mbaiwa, J.E. (2003) The socio-economic and environmental impacts of tourism development on the Okavango Delta, North-Western Botswana. *Journal of Arid Environments* 54, 447–467.

Mitchell, J. and Ashley, C. (2009) *Tourism and Poverty Reduction: Pathways to Prosperity*. London: Routledge.

Milano, C., Cheer, J. and Novelli, M. (2018) Overtourism: A growing global problem. *The Conversation Trust* (UK) http://www.theconversation.com/overtourism-a-growing-global-problem.100029

Mushongah, J. and Scoones, I. (2012) Livelihood change in rural Zimbabwe over 20 years, *Journal of Development Studies* 48(9), 1241–1257.

Nair, V. and Azmi, R. (2008) Perception of tourists on the responsible tourism concept in Langkawi, Malaysia: Are we up to it? *TEAM Journal of Hospitality & Tourism* 5(1), 27–44.

Nair, V. (2018) Lighthouse Point – responsible tourism development? *Nassau Guardian*, 28 September 2018

Nair, V., Mohamad, B., Hamzah, A., Shuib, A., Jaafar, M. and Murugesan, R.K. (2017) Multi-dimensional responsible rural tourism capacity framework for sustainable tourism. Subang Jaya, Malaysia: Centre for Research and Innovation in Tourism, Taylors University.

Nair, V., Munikrishnan, U.T., Rajaratnam, S.D. and King, N. (2014) Redefining rural tourism in Malaysia: A conceptual perspective. *Asia Pacific Journal of Tourism Research* 20 (3), 314–337.

Ngo, M. and Brklacich, M. (2014) New farmers' efforts to create a sense of place in rural communities: Insights from southern Ontario, Canada. *Agriculture and Human Values*. 31. 10.1007/s10460-013-9447-5.

Organization for Economic Co-operation and Development (OECD) (1994) *Tourism Strategies and Rural Development*. OCDE/GD (94) 49. Paris.

Pearce, D.G. (1992) Alternative tourism: Concepts, classifications, and questions. In V.L. Smith, and W.R. Eadington (eds) *Tourism Alternatives* (pp. 13–30). Philadelphia, PA: University of Pennsylvania Press.

Pugsley, P.C. (2016) *Exploring Morality and Sexuality in Asian Cinema: Cinematic Boundaries*. New York: Routledge.

Rawat, K. (2015) Exploring the role of innovation in rural areas of ASEAN for socio-economic transformation. Master of Science thesis, Taylor's University, Malaysia.

Reed, M. (1997) Power relations and community-based tourism planning. *Annals of Tourism Research* 24 (3), 566–591.

Scheyvens, R. (2002) *Tourism for Development: Empowering Communities*. Harlow, Essex: Prentice Hall.

Sharpley, R. (2000) Tourism and sustainable development: Exploring the theoretical divide. *Journal of Sustainable Tourism* 8 (1), 1–19.

Shields, B.P., Moore, S.A. and Eagles, P.F.J. (2016) Indicators for assessing good governance of protected areas: Insights from park managers in Western Australia. *PARKS*, 22 (1), 37–50.
Singh, S. (ed.) (2011) *Domestic Tourism in Asia: Diversity and Divergence*. Singapore: ISEAS.
Simkova, E. (2007) Strategic approaches to rural tourism and sustainable development of rural areas. *Agricultural Economics–Czech* 53 (6), 263–270.
Simpson, M.C. (2009) An integrated approach to assess the impacts of tourism on community development and sustainable livelihoods. *Community Development* 44 (2), 186–608.
Sorensen, A. and Nilsson, P.A. (2003) What is managed when managing rural tourism? The case of Denmark. In D. Hall, L. Roberts and M. Mitchell (eds) *New Directions in Rural Tourism*. Aldershot: Ashgate Publishing Ltd.
SRI International (2016) Wellness tourism is a growth opportunity worldwide. See https://www.sri.com/blog/wellness-tourism-growth-opportunity-worldwide (accessed 3 April 2016).
Stone, M.T. (2015) Community-based ecotourism: A collaborative partnerships perspective. *Journal of Ecotourism* 14 (2–3), 166–184. doi:http://dx.doi.org/10.1080/14724049.2015.1023309
Suansri, P. (2003) *Handbook on Community-Based Tourism*. Bangkok, Thailand: Responsible Ecological Social Tours Project.
Tosun, C. (2000) Limits to community participation in the tourism development process in developing countries. *Tourism Management* 21 (6), 613–633. DOI: 10.1016/S0261-5177(00)00009-1.
Tucker, H., Shekton, E.J. and Bae, H. (2016) Post-disaster tourism: Towards a tourism of transition. *Tourism Studies* 17 (3), https://doi.org/10.1177/1468797616671617
United Nations Environment Programme (UNEP) (2010) World Environment Day 2010. See http://www.unep.org/wed/2010 (accessed 6 June 2011).
United Nations Environment Programme (UNEP) and United Nations World Tourism Organization (UNWTO) (2005) *Making Tourism More Sustainable – A Guide for Policy Makers* (pp. 11–12).
Urry, J. and Larsen, J. (2011) *The Tourist Gaze 3.0* (3rd edn). London: Sage Publications,
Vehbia, B.O. and Doratlia, N. (2010) Assessing the impact of tourism on the physical environment of a small coastal town: Girne, Northern Cyprus. *European Planning Studies* 18 (9), 1485–1505.
Walia, S. (2019) The economic challenge of Japan's aging crisis. *The Japan Times*. Issue November 19.
Weaver, D.B. (1991) Alternative to mass tourism in Dominica. *Annals of Tourism Research* 18 (3), 414–432.
Weissmann, A. (2019) Resolving undertourism. *Travel Weekly*. January 28. See https://www.travelweekly.com/Arnie-Weissmann/Resolving-undertourism (accessed 30 March 2018).
Winter, T., Teo, P. and Chang, T.C. (eds) (2008) *Asia On Tour: Exploring the Rise of the Asian Tourist*. London, Routledge.
Wyllie, R. (1998) Not in our backyard: Opposition to tourism development in a Hawaiian community. *Tourism Recreation Research* 23 (1), 55–64.

Theme 1

Harmony Between Humans and Nature in Moulding Rural Tourism in Asia

2 Responsible Rural Tourism in Bhutan: Aligning Gross National Happiness with the Cape Town Principles

Sarah Schiffling, Chris Phelan and Karma Pema Loday

Introduction

This chapter presents the unique approach to rural tourism taken by the small Himalayan country of Bhutan. This approach, based on the national philosophy of gross national happiness (GNH), also resonates with broader conceptualizations of sustainable and responsible tourism. The chapter begins with background information on Bhutan and its tourism industry. It then goes on to discuss GNH and its impact on tourism, including case vignettes to illustrate rural tourism practices in Bhutan. The chapter concludes with reflections of lessons learned from Bhutan and how they may inform rural tourism in other developing countries.

Background

With a population of only 750,000, Bhutan is one of the smallest countries in South Asia and considerably less populous than its nearest neighbours, India to the South and the People's Republic of China to the North. Traditionally a very isolated country with a well-preserved culture, Bhutan began to open up to the outside world from the 1950s under a succession of increasingly progressive and modernizing monarchs. Bhutan joined the United Nations in 1971, welcomed the first tourists in 1974 and became a constitutional monarchy – with a democratically elected Parliament – in 2008. During this period, Bhutan established itself as one of the most stable and peaceful nations within the region (Uitz, 2012); however, it remains one of the least developed countries globally (Asian Development Bank, 2016).

Outside of the capital Thimphu, the country remains predominantly rural, with more than 70% of Bhutanese living in rural areas and 60% practicing subsistence agriculture based on a combination of crop and livestock production (Thinley & Lassoie, 2013). With such a large rural base, Bhutan inevitably experiences many of the key developmental challenges common to the Global South, especially those associated with subsistence agriculture and rural poverty. However, herein also lies its national potential, with rural Bhutan offering unique and spectacular cultural and natural attractions, and thus significant touristic appeal.

Indeed, recognizing the potential for poverty alleviation and employment generation, the Royal Government of Bhutan, in its latest 5-year plan (to 2020), identifies the tourism sector as particularly fertile for investment. Tourism is seen as a means to generate a largely absent entrepreneurial class, but also to provide much-needed economic growth. The rate of annual growth in recent years has been an enviable 6.4%. However, this obscures the fact that growth has largely been a result of investments in hydropower and electricity exports to neighbouring India. This growth is cyclical and volatile and rests on a narrow economic base. Moreover, hydropower investment provides little employment for rural communities and hence economic diversification is seen as essential (Mitra *et al.*, 2014).

Beyond its rich cultural and environmental heritage and resultant touristic appeal, Bhutan is perhaps best known for its development philosophy termed GNH. Fundamentally, GNH is a policy that assumes development should not focus solely on economic growth. Rather, it postulates that equitable socioeconomic development, environmental conservation, cultural preservation, and good governance, as interrelated concepts with underpinning Buddhist values, should be the foundations for development. Introduced in the 1970s by the country's fourth king, His Majesty Jigme Singye Wangchuck, GNH values and policies have shaped tourism policy from the moment Bhutan welcomed its first international guests. In the years since, a GNH governance framework that now consists of nine domains, 33 variables and over 120 indicators has shaped a unique approach to responsible and sustainable touristic practices (Otsubo, 2016; Schroeder, 2015).

In the sections below, this approach will be discussed in the national context, as well as in its relation to tourism development. Subsequently, opportunities for other developing countries to learn from Bhutan's approach will be highlighted.

Tourism in Bhutan

As noted above, Bhutan has been open for tourism since 1974 when the first 287 tourists travelled the country. The main objective for the introduction of tourism was the potential contribution to socioeconomic development, primarily through the facilitation of foreign exchange, but also through sharing the country's unique culture and traditions with the outside world (Dorji, 2001). By the late 1980s, tourism contributed over USD2 million in revenue to the Bhutanese economy, despite low tourist numbers. By 1992, tourist revenue of USD3.3 million accounted for 15–20% of Bhutan's exported goods and services. Since then, the sector has seen considerable investment and steady growth. In 2017, a record 254,704 visitors arrived in Bhutan, mostly from countries within the region (with 183,287 from India, Bangladesh and the Maldives). Of the total, 71,417 were considered international, thus non-regional arrivals. This represents a growth of 21.5% on the previous year (Tourism Council of Bhutan, 2017).

Despite this growth, Bhutan is aiming for a selective tourism market, exploiting its exclusivity as a travel destination and its reputation for authenticity, remoteness and a well-protected cultural heritage and natural environment. From the outset, the mantra of Bhutanese tourism was 'high value, low volume'. From the 1970s onwards, the prevailing view was that uncontrolled and unrestricted tourism would threaten Bhutan's natural environment and culture heritage, as well as overburden its limited infrastructure. The high-value/low-volume policy was established on the basis of a quota on the number of tourists allowed entry to Bhutan each year. Moreover, visitors were required to pay a daily tariff of USD130 and to join a package tour of six or more people. In the early decades, these tours were wholly government-owned and operated. However, from the 1990s – and with the desire to introduce some competition – the high-value/low-volume mantra was subtly substituted for 'high value, low impact'. At the same time, the quota ended, but the daily tariff was increased to USD200 in the high season and USD165 in the low season. Today, it stands at USD250/USD200. While initial quotas sought to limit potential cultural and environmental degradation, the removal of the tariff, but an increase in the daily rate, was designed to maximize revenue, while still restricting numbers. This was founded on the belief that only those willing to pay the high tariff for unique experiences would be attracted to Bhutan (Dorji, 2001; Schroeder & Sproule-Jones, 2012; Suntikul & Dorji, 2016).

Even today, international tourists cannot travel independently in Bhutan, but are required to be a part of guided package tours, which are highly customisable and are conducted in small groups or even for individual travellers. Bhutan places no such restrictions on regional visitors from India, Bangladesh and the Maldives. Capacity restrictions and tariffs relate only to non-regional visits. To see numbers as the prime sustainability measure is arguably naive. The argument presented is that GNH, as a governance framework, is key. With the emphasis on high value, low impact, the importance of responsible tourism as a contributing factor to socio-economic development is recognized. Under a GNH governance framework, sustainability is at the core in Bhutan. Rather than relying on foreign investment and concerned about leakages, the development of a truly indigenous industry has been prioritised.

Outside the capital city Thimphu, Bhutan is almost exclusively rural with predominantly minor agricultural settlements and some small market towns. According to the Tourism Council of Bhutan (2017), together the districts of Thimphu and Paro, where the sole international airport is located, account for over 60% of bed nights. The top four most visited dzongkhags (districts) made up 85% of bed nights in 2017, while many of the remaining 16 districts see only extremely limited tourism. The impact is therefore quite concentrated on a small area of the very sensitive natural environment, which further adds to the imperative of emphasizing sustainable rural tourism development.

On average, tourists spend one week in Bhutan and almost 90% of them claim the primary motivation for their visit is cultural. However, there is also a distinct form of tourism that focuses on the natural environment and explores some of the more remote areas of Bhutan. In 2017, 4354 tourists were recorded on the 26 most popular trekking routes. As with the dzongkhags visited, it is notable that there is a concentration on a few well-frequented trekking routes, with the top two alone accounting for 48% of trekkers. The most popular is the Druk Path Trek, a five-day trek along an ancient high-level route between Paro and Thimphu. Given the unique geography of Bhutan with its high mountains and narrow valleys even a trek in such a central part of the country leads through remote rural areas (Tourism Council of Bhutan, 2017).

Gross National Happiness

As outlined above, Bhutan has become synonymous with the development policy of GNH. From as early as the 1960s, the Bhutanese government adopted a five-year planning system as its approach to implementing development policies and programmes, emphasizing the development of human resources, preservation and promotion of cultural and traditional values, people's participation and private sector development (Planning Commission, 1996). These gradually helped to build the national development philosophy known as 'gross national happiness'. GNH was conceived as an alternative to the focus on gross domestic product (GDP) in dominant development approaches. GNH is a distinctly Bhutanese concept first articulated by His Majesty Jigme Singye Wangchuck in the 1980s. While definitions vary, a commonly accepted one states that GNH 'measures the quality of a country in a more holistic way and believes that the beneficial development of human society takes place when material and spiritual development occurs side by side to complement and reinforce each other' (Otsubo, 2016: 269). A key difference to conventional development approaches is the strong focus on living in harmony with nature and with respect for others, recognizing that each individual has material, but also spiritual and emotional needs. Essentially this alternative development approach mirrors the Asian philosophy of harmony between humans and nature (discussed in Chapter 1) that has shaped the sustainable growth of a predominantly rural form of tourism in Bhutan.

Bhutan pursues developmental activities based on GNH concepts by identifying the national goals under four broad key strategies. Popularly known as the four GNH pillars, these are:

(1) sustainable and equitable socioeconomic development;
(2) conservation of environment;
(3) preservation and promotion of culture;
(4) promotion of good governance.

The four pillars are further measured by a GNH index, which is a composite statistic of measuring Bhutan's progress in enhancing the happiness of the people based on a holistic framework made up of nine domains and 33 indicators (Gross National Happiness Commission, 2012). The nine domains are:

(1) psychological well-being;
(2) health;
(3) time use;
(4) education;
(5) cultural diversity and resilience;
(6) good governance;
(7) community vitality;
(8) ecological diversity and resilience;
(9) living standard.

Sustainable Development and Responsible Tourism

Like many developing nations, Bhutan identified tourism as an integral aspect of their economic development strategy; while remaining keenly aware of the potentially negative and destructive consequences of tourism development on Bhutanese environment and society (Sharpley, 2009). From the discussion above, it should be no surprise, therefore, that a national tourism strategy of high value, low volume (later evolving to low impact) would emerge from a governance framework and set of values such as GNH.

For Brunet *et al.* (2001), the demand for authentic culture- and nature-based tourism has been both a blessing and a curse for Bhutan. Bhutan wishes to share in the economic development bounty enjoyed by many small developing countries, while being keenly aware of the challenges that increased tourism can bring. Certainly, the development trajectory for Bhutan could have been very different. Schroder and Sproule-Jones (2012) contrast the tourism policies of Nepal and Bhutan, locations with geographic similarities and both offering unique and distinct cultural and natural assets to the visitor, yet whose development pathways could not have been more different. They conclude that Nepal has been driven primarily by motivations of economic self-interest among key stakeholders, whereas Bhutan has prioritised cultural and environmental preservation over economic factors (Nepal & Karst, 2017). That is not to say that economic motivations are not apparent, with Sharpley (2009) reminding us that the high daily tariff – while restricting mass tourism – actually promotes a higher spend from an arguably higher class of visitors. Indeed, for Sharpley, Bhutan now has a 'scarcity value' that is in itself an asset.

Certainly, the high-value and low-volume/impact story has been reported on widely in the tourism literature (Dorji, 2001; Schroeder, 2015), although as Nyaupane and Timothy (2010) highlight, this policy is not applied universally. Specifically, Bhutan's controlled tourism policy is only applied to western tourists and not to those from regional neighbours such as India, Bangladesh and the Maldives. Uncontrolled numbers from India, in particular, give rise to carrying capacity concerns, which might suggest that Bhutan's tourism policy of advocating environmental and cultural protection is being challenged by regional politics.

The Cape Town Declaration, Responsible Tourism and GNH Values

While acknowledging the concern that the Bhutanese approach may be driven as much by isolationism and realpolitik (Nyaupane & Timothy, 2010), it is nonetheless the case that considerable attention has been paid to the sustainability aspects of Bhutanese tourism policy. This has been considered through the lens of ecotourism (Gurung & Scholz, 2008; Karst, 2017; Rinzin *et al.*, 2007), responsible tourism (Hummel *et al.*, 2013), or sustainable tourism (Dorji, 2001; Schroeder, 2015; Sharpley, 2009). Indeed, the varying descriptors of ecotourism, responsible tourism and sustainable tourism, applied to Bhutan, reflect the myriad of terms used – often interchangeably – within the broader academic and policy literature. Indeed, within wider tourism discourse, the list might be extended to include discussions of the terms geotourism, cultural tourism and pro-poor tourism (Edgell, 2016).

Definitions of sustainable tourism tend to emerge from the core principles of sustainable development established by the Brundtland Report (1987), that sustainability is about meeting the needs of the present without compromising the ability of future generations to meet their own needs. However, while sustainable tourism is often equated with triple bottom-line definitions of economic, environmental and sociocultural sustainability, Sharpley (2009) highlights that we have yet to achieve definitional consensus, despite decades of academic attention. For purposes of this chapter, the more recent conceptualization of 'responsible tourism' is adopted. Responsibility in this context means an articulation of tourism that aims to minimize negative social and environmental impact, while maintaining the benefits for local communities, and allowing local people to conserve fragile cultures, habitats and species (Edgell, 2016; Goodwin, 2016). The local in this context is important and indeed the responsible tourism agenda is said to be broader, placing 'quality of life at its core' and asking 'what individuals and groups do to address sustainability issues which arise in particular places' (Goodwin, 2016: 17).

The Cape Town Conference on Responsible Tourism in Destinations was held in 2002. Building on existing South African work on responsible tourism, it resulted in what has become known as the 'Cape Town Declaration'. This declaration recognizes the importance of the World Tourism Organization's Global Code of Ethics, which aims to promote responsible, sustainable and universally accessible tourism. Furthermore, it is committed to equitable, responsible and sustainable world tourism and is supportive of the World Tourism Organization's efforts to harness sustainable tourism to help eliminate poverty (Goodwin, 2016).

The following sections will trace the seven elements of the Cape Town Declaration in the Bhutanese tourism sector, further expanding upon the alignment between the declaration and the GNH policy framework. To illustrate tourism practice, reference will be made to a variety of primary and secondary research in the form of case vignettes. To aid the reader, Table 2.1 demonstrates how the Cape Town principles and GNH values (in the context of tourism) might align.

Minimizing Impacts

The first of the Cape Town principles emphasizes the minimization of negative economic, environmental and social impacts. Here, the high daily tariff is purposely designed to limit environmental and social impact, while maintaining economic gains. Moreover, in the early years, it specifically mitigated against ecological concerns. In more recent years, proposals to liberalize tourism policy and increase tourist numbers has given rise to concerns that the cultural heritage and environment will suffer; with one tourism operator in a study by Schroeder (2015) stating that it, 'frightens me', adding that 'as Bhutanese, we feel so much for preservation of culture and identity. If we liberalize, we'll be no different from Nepal'.

Regarding specific niches, while cultural visits are the highest recorded category, the adventure tourism market consists of just under 10% of all arrivals. Within this percentage the breakdown of activities like hiking, trekking, kayaking, mountain biking and fishing is not known, though as has already been noted, 4534 tourists walked Bhutanese trekking routes in 2017 (Tourism Council of Bhutan, 2017). Therefore, the potential impact of tourism on the remote rural and mountainous ecosystem is

Table 2.1 The Cape Town Declaration on Responsible Tourism, mapped against GNH pillars

Gross National Happiness Pillars → Cape Town Responsible Tourism Principles ↓	Sustainable and equitable socioeconomic development	Conservation of environment	Preservation and promotion of culture	Promotion of good governance
Minimizes negative economic, environmental, and social impacts	x	x	x	
Generates greater economic benefits for local people and enhances the well-being of host communities, improves working conditions and access to the industry	x			x
Involves local people in decisions that affect their lives and life chances				x
Makes positive contributions to the conservation of natural and cultural heritage, to the maintenance of the world's diversity		X	x	
Provides more enjoyable experiences for tourists through more meaningful connections with local people, and a greater understanding of local cultural, social and environmental issues		X	x	
Provides access for physically challenged people;				
Is culturally sensitive, engenders respect between tourists and hosts, and builds local pride and confidence.	x	X		

critical, with adventure companies legally mandated to camp only in designated areas, to register with the caretaker of the site, pay site fees, and carry out all refuse. Nonetheless, complaints about extreme littering on campsites on the most popular treks are not uncommon (Horrell, 2011). To address this, Clean Bhutan (2018) conducts regular clean-up campaigns along trekking routes, as well as in towns and along rivers, recognizing a garbage problem that has developed with the rapid economic growth of the country, including the growth of tourism.

More broadly, the Bhutanese constitution establishes the need to maintain a minimum of 60% of total land area under forest cover now and into the future. In 2017, the national audit confirmed that forest cover is currently at 71%. Through the mandated use of kerosene rather than firewood, the trekking industry has managed to avoid contributing to deforestation (WWF, 2018).

Many of Bhutan's iconic Himalayan peaks remain unclimbed or have not been scaled in decades. Bhutan has the highest unclimbed mountain in the world, 7570 m Gangkar Punsum. This is due to social concerns as many of the highest mountains are considered sacred by the local population (Karst, 2017). While other countries, such as Nepal, have confronted the problem of offending the mountain gods by a permit system and designated holy mountains which remain unclimbed, Bhutan chose to introduce a blanket ban. It is prohibited to climb any peak above 6000 m in Bhutan. This shows the strong link of responsible tourism practice and the GNH philosophy, underpinned by religious beliefs and cultural norms.

With regard to economic impact, tourism in Bhutan generates greater economic benefits for local people, enhances the well-being of host communities and improves working conditions and access to the industry. Tour operators from Thimphu stress that tourism operations bring a range of employment opportunities to the rural communities they frequent, from small road-side vendors to large resorts. As there are strict rules for foreign investment and the employment of foreign nationals, operations of all sizes are staffed and owned by Bhutanese citizens.

Community Involvement

Involving local people in decisions that affect their lives and life chances is the second Cape Town principle. Bhutanese tourism was very much centrally planned and implemented in the early years. As the industry matured, both government and local communities are now arguing for stronger community involvement to further rural development. While tourists visit rural areas, many guides and tourism entrepreneurs are from the capital region. In many cases, local community involvement is limited to providing yak or horse drivers to transport luggage. To change this, community-based tourism is encouraged, for example, in the Jigme Dorji National Park. It is envisaged that the participation of local communities in tourism activities will spread their financial benefits and also enhance the skills base and buy-in necessary to further the national tourism development (Gurung & Scholz, 2008; WWF, 2018).

River Guides of Panbang is a rafting company operating in the Royal Manas National Park. They demonstrate a strong connection with the community that is common among tourism operators: bringing tourists to a remote region that depends primarily on subsistence farming; providing villagers with an opportunity to sell their traditional craft items. In addition, the operator partners with Clean Bhutan to coordinate clean-up campaigns. Furthermore, the business donates 5% of its earnings to the Royal Manas National Park and 20% to the Panbang community (Phelan et al., 2017).

Conservation of Natural and Cultural Heritage

While the discussion above has already emphasized the positive aspects of environmental and sociocultural preservation, the Cape Town principles specifically highlight the need for responsible tourism to make positive contributions to the conservation of natural and cultural heritage and to the maintenance of the world's diversity.

The Royal Society for the Preservation of Nature (2016) has the vision that future generations of Bhutan will live in an environmentally sustainable society. Among the projects they champion is a community-based tourism model in the Phobjikha Valley that contributes to conservation of the natural environment, conserves local culture and contributes to socio-economic benefits for local communities. The valley is one of the major habitats of the black-necked cranes in Bhutan, a bird sacred in Bhutanese culture, often appearing in folklore, dances and historical texts.

To achieve this, they generate awareness among tour guides and locals on the importance of wildlife viewing as an aspect of the local tourism offer. Moreover, they enhance alternative income opportunities for local communities by supporting tourism-related skills development and income diversification activities. This encourages a reduction in human encroachment in the wetland and threats to the cranes from pesticides, as subsistence farmers begin to realize the touristic and ecosystems value of the cranes. Further developments have seen the establishment of the Black-Necked Crane Festival, with the preservation of the natural heritage offering synergistic benefits also for regional cultural heritage.

Regarding cultural heritage, while Western fashion is influencing Bhutan, traditional dress is an obvious and striking aspect of the culture to any visitor. Men wear the Gho, a knee-length robe and women wear the Kira, a long, ankle-length dress accompanied by a light outer jacket. Bhutanese wear long scarves when visiting administrative centres. The scarves worn vary in colour, signifying the wearer's status or rank. Traditional food is ever-present, even though some hotels will serve Western food or other Asian cuisines as well. As an example, taxi drivers are mandated to wear traditional dress at all times and in almost all service encounters, tourists will have similar experiences. Traditional dress is seen to be essential to protect and nurture Bhutan's culture to help guard the sovereignty of the nation. In the 2017 tourism report, Bhutanese culture features as the key reason for visiting, with religious reasons also stated. Additional key sights in Bhutan are monasteries and fortresses, both intricately tied to the unique culture and history of the country (Tourism Council of Bhutan, 2007).

The Challenge of Accessibility

Providing access for physically challenged individuals is the fourth responsible tourism principle and as Table 2.1 highlights, is the key Cape Town objective that does not readily align with GNH. However, this is unsurprising as, like many of the world's least developed countries, Bhutan has an infrastructure challenge. Travelling around Bhutan with a physical disability can be very difficult. A pavement or sidewalk is rare and where they do exist, are often in a poor state of repair. Access ramps and lifts are seldom provided in hotels, and the same is true for disabled/accessible toilet facilities. What is more, the rural landscape of steep mountainsides often traversed by rough pathways can make many key sights, such as the iconic Taktshang Goemba (Tiger's Nest Monastery), inaccessible to all but the physically fit. Guidebooks like the current edition of the *Lonely Planet* (Mayhew & Brown, 2017) highlight that 'a cultural tour in Bhutan is a challenge for a traveller with physical disabilities, but is possible with some planning'. However, they point out that the willingness of locals to provide assistance is immense.

Sensitivity, Pride and Respect

In an obvious departure from traditional sustainability definitions, the Cape Town declaration emphasizes as its final principle that tourism should be culturally sensitive, engender respect between tourists and hosts, and build local pride and confidence. To this end, pride in the country's unique culture and history is strong within Bhutan and tourists are encouraged to explore this. Cultural preservation as a key tenet of GNH is taken seriously by all, but at the same time Bhutan is also open to outside influences. As everywhere else, Bhutanese youth are interacting globally on social media. Many also travel abroad for their education. All this gives rise to a concern over dilution of Bhutanese traditions. However, this provides a key enabling role for the tourism industry, where modern income-generating activities and skilled jobs can be tied to culture heritage. Tourists to Bhutan travel primarily for culture, thus showing the importance and offering additional reasons for pride in the national culture. Overall, local pride and confidence are very closely associated with tourism in Bhutan. There is a great respect for guests and Bhutan is a very safe country to travel to, which further adds to the positive tourism experience. While Bhutan is not the untouched Shangri-La, it is sometimes portrayed as such; tourism in the country is built strongly on culturally sensitive relationships between tourists and host.

Conclusion

As the case above identifies, there is a synergy between the GNH pillars and the Cape Town principles of responsible tourism. Like many destinations, Bhutan recognizes tourism as a driver to address development challenges, to diversify its rural economy and alleviate poverty. In doing so, it prioritises responsibility for the environment, culture and communities, over purely economic gain. The spiritual reverence for sacred natural sites especially the mountains has been translated into a national policy of prohibiting tourists from climbing any peak above 6000 m in Bhutan. Implicitly this approach reflects a strong alignment with the Asian philosophy of harmony between humans and nature that has been shaped by beliefs, taboos and reverence. Goodwin (2016) reminds us that there are places that use tourism, while others are used by it. This case illustrates that Bhutan is presently a 'user' and not 'the used'. Rural tourism brings evident benefits to Bhutan and arguably the high-value, low-impact approach has been successful without sacrificing the spirituality of its rural areas. The comparisons with Nepal are appropriate, given that both have placed tourism central to development, while framing this very differently in policy and governance terms.

In respect to other countries, particularly those of the Global South, the Bhutan case highlights that the principles of responsible tourism can be enshrined in the national tourism policy. Strong governance frameworks can ensure that tourism development occurs in a manner consistent with the Cape Town declaration, that is, in a way that minimizes negative social and environmental impact, maintains benefits for local communities and maintains (if not rejuvenates) fragile cultures. What remains unclear and where further research can elaborate, is the extent to which these ideals are bottom-up or top-down. GNH is not entirely autocratic. As the case vignettes illustrate, tourism operators and communities evidently have significant implementation responsibilities. What balance is necessary and what further lessons can GNH offer for responsible rural tourism initiatives elsewhere?

However, one must question if the approach can sustain Bhutan going forward. Tourist arrivals have steadily increased and while GNH values have ensured an appropriate development trajectory thus far, a tipping point could soon be reached; social and environmental carrying capacity must remain key considerations. Certainly, the criticism that Bhutanese tourism policy may be as much realpolitik, given that visitors from neighbouring India are unrestricted, as much as a philosophical alignment with sustainable development policies, warrants further scrutiny. Nonetheless, to conclude, Bhutan offers a rich learning opportunity for the responsible tourism movement, though equally it can itself continue to learn from the hard lessons experienced by destinations elsewhere.

References

Asian Development Bank (2016) *Key Indicators for Asia and Pacific 2016*. Manila. See https://www.adb.org/sites/default/files/publication/ 204091/ki2016.pdf (accessed 12 January 2018).
Brundtland, G. (1987) *Report of the World Commission on Environment and Development: Our Common future*. See http://www.un-documents.net/our-common-future.pdf (accessed 11 September 2018).
Brunet, S., Bauer, J., De Lacy, T. and Tshering, K. (2001) Tourism development in Bhutan: Tensions between tradition and modernity. *Journal of Sustainable Tourism* 9 (3), 243–263.
Clean Bhutan (2018) *Taktsang Trail Clean-up Program*. See https://cleanbhutan. org/?page_id=2844 (accessed 28 March 2019).
Dorji, T. (2001) Sustainability of tourism in Bhutan. *Journal of Bhutan Studies* 3 (1), 84–104.
Edgell, D. (2016) *Managing Sustainable Tourism: A Legacy for the Future*. Abingdon: Routledge.
Gross National Happiness Commission (2012) Eleventh Five Year Plan 2013–2018. Thimphu: Royal Government of Bhutan.
Goodwin, H. (2016) *Responsible Tourism*. Oxford: Goodfellow.
Gurung, D.B. and Scholz, R.W. (2008) Community-based ecotourism in Bhutan: Expert evaluation of stakeholder-based scenarios. *The International Journal of Sustainable Development and World Ecology* 15 (5), 397–411.
Horrell, M. (2011) *Yakking with the Thunder Dragon*. Los Gatos, CA: Smashwords.
Hummel, J., Gujadhur, T. and Ritsma, N. (2013) Evolution of tourism approaches for poverty reduction impact in SNV Asia: Cases from Lao PDR, Bhutan and Vietnam. *Asia Pacific Journal of Tourism Research* 18 (4), 369–384.
Karst, H. (2017) This is a holy place of Ama Jomo: Buen vivir, indigenous voices and ecotourism development in a protected area of Bhutan. *Journal of Sustainable Tourism* 25 (6), 746–762.
Mayhew, B. and Brown, L. (2017) *Lonely Planet Bhutan*. Singapore: Lonely Planet Global Limited.
Mitra, S., Carrington, S. and Baluga, A. (2014) Unlocking Bhutan's potential: Measuring potential output for the small, landlocked Himalayan Kingdom of Bhutan. See https://www.adb.org/sites/default/files/publication/152571/south-asia-wp-032.pdf (accessed 15 March 2017).
Nepal, S. and Karst, H. (2017) Tourism in Bhutan and Nepal. In C.M. Hall and S.J. Page (eds) *The Routledge Handbook of Tourism in Asia*. Abingdon: Routledge
Nyapune, G.P. and Timothy, D.J. (2010) Power, regionalism and tourism policy in Bhutan. *Annals of Tourism Research* 37 (4), 969–988.
Otsubo, S.T. (2016) *Globalization and development – In Search of a New Development Paradigm*. Abingdon: Routledge
Phelan, C., Loday. K.P. and Schiffling, S. (2017) Culturally sustainable entrepreneurship: Realising socio-economic development in Bhutan. Paper presented at the *15th Rural Entrepreneurship Conference*, 15–16 June 2017, Newcastle University, UK.
Planning Commission (1996) Eighth Five Year Plan (Vol. 1). Thimphu: Royal Government of Bhutan.

Royal Society for Protection of Nature (2016) Phobjikha Conservation Area. See http://www.rspnbhutan.org/phobjikha-conservation-area/ (accessed 17 August 2018).

Rinzin, C., Vermeulen, W.J.V. and Glasbergen, P. (2007) Ecotourism as a mechanism for sustainable development: The case of Bhutan. *Environmental Sciences* 4 (2), 109–125

Schroeder, K. and Sproule-Jones, M. (2012) Culture and policies for sustainable tourism: A South Asian comparison. *Journal of Comparative Policy Analysis: Research and Practice* 14 (4), 330–351.

Schroeder, K. (2015) Cultural values and sustainable tourism governance in Bhutan. *Sustainability* 7 (12), 16616–16630.

Sharpley, R. (2009) *Tourism Development and the Environment: Beyond Sustainability?* Earthscan: London.

Suntikul, W. and Dorji, U. (2016) Tourism development: The challenges of achieving sustainable livelihoods in Bhutan's remote reaches. *International Journal of Tourism Research* 8 (5), 447–457.

Tourism Council of Bhutan (2017) Bhutan Tourism Monitor 2017. See https://www.tourism.gov.bt/uploads/attachment_files/tcb_buHnrvHE_BTM%202017.pdf (accessed 30 December 2018).

Thinley, P. and Lassoie, J.P. (2013) Promoting biodiversity conservation and rural livelihoods in Bhutan. See http://www.conservationbridge.org/wp-content/media/2011/12/Promoting-Biodiversity-Conservation-and-Rural-Livelihoods-in-Bhutan.pdf (accessed 23 May 2018).

Uitz, M. (2012) *Hidden Bhutan: Entering the Kingdom of the Thunder Dragon*. London: Haus Publishing.

WWF (2018) Bhutan: Committed to conservation. See https://www.worldwildlife.org/projects/bhutan-committed-to-conservation (accessed 29 July 2019).

3 Crown Cave, Guilin: A Chinese Perspective on Responsible Rural Tourism

Trevor H.B. Sofield and Fung Mei Sarah Li

Introduction

In order to understand the approach adopted by Chinese authorities in developing and presenting caves for tourism, it is necessary first of all to look briefly at the western underpinnings of caves being used as a tourism resource. The differences are stark. In the West, the biocentric approach is the dominant paradigm. This western perspective separates nature and civilization (humans). Biocentricity emphasizes the retention of naturalness, if necessary at the expense of recreational and other human uses. The principle of conservation ideally takes primacy over economic profit-making and human comfort. In China, the 4000-year-old Daoist tenet of 'humans in harmony with nature' guides society's value system (Xu *et al.*, 2014).

Thus caves are managed anthropocentrically as cultural rather than natural sites, with many artificial factors added to the natural characteristics. Since humans and nature are indivisible, 'artificiality' made by humans to 'improve' nature is 'natural'. Western countries tend to draw a strong distinction between cultural heritage caves (those that hold archaeological value related to ancient human settlement/use, or as religious sites or defence fortifications) and natural caves undisturbed by human hand, and manage them according to the dominant discipline involved in their conservation and protection. However, in China, rarely is such a distinction drawn and all caves are treated in a culturally determined manner. Under these circumstances, the concept of 'responsible rural tourism' has very different overtones if viewed through a narrow conservation lens, and socio-ecological definitions as to what constitutes 'responsible' are also dissimilar. In itself, difference does not necessarily make one approach right and the other wrong, and in this chapter we explore these differences.

Western Management of Caves

Most western countries have enacted legislation to protect the ecology of caves. For example, the US Forest Service compiled regulations for implementing the Federal Cave Resource Protection Act in 1994 which determines environmentally sound management practices for caves in the United States (Stitt, 1994). In 2003, New Zealand produced a comprehensive handbook on the sustainable management of natural assets used for tourism, which incorporated a detailed section on management guidelines, indicators and monitoring of tourism visitation for caves (Hughey & Ward, 2003). A cave classification system adopted (with minor variations) by most western countries constitutes the foundation of the western paradigm as it concerns the management of caves (see Table 3.1).

It is clear from this classification system that the biocentric approach predominates and is focused on the conservation ethic regarding all tourism to a site and minimizing environmental impacts to the greatest possible extent (Gillieson, 1996). For example, lighting will be designed to be as natural as possible (no brightly coloured lights) to minimize disturbance to the specialized life forms that have adapted to the low light of caves (e.g. bats, glow-worms, a variety of insects such as albino cockroaches and blind earwigs, transparent fish, etc.). Flash photography may also be banned, especially if there is wildlife (e.g. glow-worms) sensitive to light. Similarly, noise levels may be monitored in some caves. Visitation will be restricted, even banned, during periods of sensitivity (e.g. breeding season of bats). A 'Look but do not touch' policy will be implemented in order to protect and conserve delicate formations ('decorations' is the technical term). Interpretation will be scientific, based on explaining the geology of the caves and the chemical processes leading to different formations and the highly specialized habitats and lifecycles of the various animals that live in the particular habitats of an underground environment (Gillieson, 1996; Ham, 1992).

Education is the principle objective. Making a presentation interesting is obviously necessary to achieve the best educational outcome but entertainment as such is subordinated to the pedagogy of environmental 'best practice'. Examples of 'biocentricity first' include the Undaralava tube caves in Queensland, Australia that are closed for 10 weeks each year during the bat breeding season; and the world-famous Waitomo Glowworm Caves in New Zealand where visitors are very strictly monitored and controlled to minimize impact on the glow-worms. No flash photography is allowed and noise must be minimized; carbon dioxide levels are monitored hourly and if they rise too high, tourists will be evacuated and the caves will be closed until the concentration drops to an acceptable intensity. About 30 years ago, the caves were closed for a year to allow glow-worm populations to recover from tourism-related disturbance that restricted their breeding and saw a population collapse of more than 50%.

Table 3.1 Caves classification system (after Gillieson, 1996)

GROUP	CLASSIFICATION	DESCRIPTION
1	Closed Caves	Access not permitted because of danger from instability or foul air, or – Caves awaiting classification.
2	Scientific Reference Caves	Caves which are best representatives of particular attributes of geology, geomorphology, biology or archaeology, where 'the management aim will be to preserve the caves in their natural state so that reference sets of caves & cave life are available in perpetuity' (Gillieson 1996: 254). Access will be restricted to valid scientific research purposes under strict controls.
3	Limited Access Caves	The quality of their physical or biological attributes merits special protection; or the level of physical difficulty in entering and exploring them must be limited to experienced speleologists only. Visitor group size will be strictly limited, and scientific research will be encouraged. Gating will be utilized to assist management.
4	Speleological Access Caves	The physical and biological attributes do not require special protection but the level of physical difficulty in exploring the cave will provide a good range of recreational opportunities for speleologists. Gating is desirable and again, group sizes will be strictly limited.
5	Adventure Caves	No particular inherent value or significance other than their morphology which will make them suitable for exploration by relatively inexperienced cavers, or youth groups (under supervision), or for training, or perhaps providing a 'soft' speleological experience for tourists.
6	Public Access Caves	Those which are opened for general tourism/visitation/recreation either with guides or on unescorted tours. Interpretation will be 'deep scientific', based on explaining geology, the chemical and physical processes that have produced the formations inside the cave, and the biology of organisms (plant & animal life) inside the cave environment.

Today the Waitomo Caves complex operates under a Scientific Advisory Group to ensure that environmental best practice standards are maintained.

Chinese Paradigm – A Contrast

In China, caves as contemporary tourist attractions are embedded in millennia of culturally derived and culturally determined values that remain valid today. The result is that in China cave tourism is overwhelmingly a cultural experience in a natural site.

For centuries, caves have been central aspects of Daoist belief/philosophy, as Confucian power places for meditating, gaining inner strength and self-knowledge, as sacred sites for Buddhist worship and veneration, as special healing places, and as cultural sites reflecting classical literature, ancient poems, calligraphy (high art), and other literary values (Li, 2006). Caves as a particularly important component of Daoism are 'gateways to Heaven' (*dong*) and since ancient times there have been 36 nationally famous sacred Daoist pilgrimage cave sites all over China, with many hundreds more having a local level of significance. As such, caves appear in many of the Chinese classics, and the histories and folklore of 2000–3000 years of continuous civilization place them in a prominent position.

Typically, the Daoists built temple across the entrance to the cave emphasizing their role as a gateway, such as the White Horse Cave near Xilin Gorge, Yichang, Hubei Province. It takes its name from a formation which looks like the steed that carried Tang Seng from China to India. When the journey was completed, the white horse rested inside this cave, so the story goes. It has remained there ever since and about 1000 years ago, a Daoist temple was constructed across its entrance. For Buddhists, caves are equally important as cultural and religious sites. Nan Tan Cave in the famous Daoist mountains of Wudangshan, has been a 'power place' for centuries and used for meditation and living quarters, a special place for sacred artefacts and objects of veneration, a library for ancient texts, with a 1200-year-old temple constructed across its entrance (Li, 2006).

Relational and Contextual Values about Nature and Naming of Caves

Daoism views humans and nature as indivisible (Chan, 1969). Chinese constructs of nature are therefore relational and contextual, and this induction of 'relational' transforms natural sites such as caves into cultural sites. For example, westerners may label an object, 'Dog Rock' because of its resemblance to a canine. But the Chinese will place that rock in a relational context and it will carry the name of 'The Dog Guarding the Two Fairies' because it faces (looks towards) nearby twin peaks, and in Chinese myths and legends, peaks are the abode of fairies. Westerners might simply name the mountains 'Twin Peaks'. The Chinese will label the peaks as, 'Two Fairies Whispering a Secret about the Monkey King'; because the Chinese will also anthropomorphize the 'mysterious mist-shrouded peaks' and insert a human activity to capture that perception of 'mysterious'. 'The Monkey King' will refer to a large cave that can be seen from the peaks which Chinese will call 'The Cave of the Monkey King' because it has a waterfall flowing down in front and this mythical creature inhabited such a cave. Westerners would simply call it 'Waterfall Cave'. The reference to the Monkey King is a theme common in caves all over China, based on a famous 16th century classic story known to all literate Chinese called 'Journey to the West'. It relates the story of a young Buddhist monk, Tang Seng, who together with the Monkey King and a host of other characters made the journey to India in the 7th century to bring back to China the Buddhist texts. Eighty out of 100 chapters take place in caves. The naming of physical natural objects illustrates the contextual and relational characteristics of the Chinese view of nature. They are not seen in isolation but as part of a complex holistic landscape; they are often imbued with human feelings and sentiments. Hence the naming of the Monkey King's cave can only be understood in terms of its contextual relationship to the entire landscape and to humans (Li, 2006).

While in the past, the main Chinese use of caves may have been religious in form and motivation, the re-use of caves today as touristic sites retains the ancient cultural values and allows contemporary Chinese to span the centuries of their rich past since they are in effect ancient heritage sites and not natural sites where visitation is governed by environmental conservation values. Pilgrimage was banned from 1949 under Mao Zedong, tourism was viewed as a bourgeois activity and thus all visitation to caves – and virtually all other scenic sites and attractions around China – ceased (Sofield & Li, 1998). After 1978 when tourism became an accepted form of economic development, many caves around China were opened up for touristic visitation and in addition to those with known religious, heritage, historical and cultural credentials, there are now more than 200 new sites as well. However, in developing them for tourism, the perception of caves as cultural sites is the foundation upon which the interpretation is based even when there has been no longstanding heritage associated with them (Gan, 1988). The expert called in to write the script for the interpretation and training of guides will often be a historian or classicist, not a geologist or biologist. A theme, or series of themes from history, religion, a literary classic, or myths and legends, will be selected around which to weave the guided tour. Individual geological features will be likened to and labelled after famous characters (human and non-human) from history, myth, legend, folklore, religion and literary texts.

For example: 'Here is the monkey king fighting the dragon king of the East Sea, and look! There are three Immortals waging bets on who will win the battle!' 'Here is the Immortal Peach being taken to the Feast of Everlasting Existence by the Jade Emperor and his wife Xi Wangmu (Queen Mother of the West)' – a guide's interpretation of Changsheng Cave, Wufeng County, Hubei Province (Li, 2006). There is no explanation provided because Chinese are familiar with all of these characters and objects and the commentary will be incomprehensible to a westerner not educated in the culture of China. Most tourist caves are described as 'fairylands' or 'wonderlands'. But the Daoist concept of 'Heaven' or 'wonderland' or 'fairyland' which is synonymous with caves is very different from a western understanding. The traditional Chinese concept of a cave 'Fairyland' is a world inhabited by 'Immortals' who carouse, play tricks, live the good life. It is not 'Heaven' in the western Christian sense, nor a place of serenity as in Buddhism's Nirvana. Nor is it a childhood fantasy, but for some Chinese, a serious alternative to a hell.

Since Confucianism, aligned with Daoism, invokes a responsibility to bring imperfect nature into harmony with humans, a common feature in many caves, will be 'improvements' of the natural cave-scapes. It is not necessarily a belief in Confucian thought but it is in the Confucian tradition of 'Man improving on Nature'. Taking a broken stalactite and turning it into a fountain in a pool and surrounding it with a circle of white pebbles and a bracelet of twinkling red lights is an example of Man improving on Nature to bring Nature into harmony with Man (Hongpin Cave in Hubei Province). Without the stalactite fountain, there is no focal point of the pool for the tourist gaze; the empty cavern has no relational or contextual values for most Chinese tourists. But the artificial addition immediately relates them to their ancient Daoist and Confucian belief systems, it is 'good *feng-shui*', and provides a context that they understand, while at the same time the perception of a 'boring' place is replaced with an entertaining sight (Li, 2006). Throughout caves in China, all kinds of additions will be constructed to make them more appealing to Chinese visitors and increase their entertainment value. They range from physical constructions like new temples and statues of Buddha, statues of Daoist gods and the 12 Immortals inside caves, inscribing calligraphy (often poems) on cave walls, to sound-and-light shows and performances of all kinds, and a wide range of photographic opportunities where visitors can climb into and onto formations for special images to be taken, all supported by souvenir and refreshment stalls inside the caves.

In a one-hour commentary, there may be two minutes of geological explanation. There is usually no biological interpretation because the activities of lights, colour, sound and action will have destroyed or driven virtually all living organisms from the cave. In visiting caves, many Chinese expect to see beautiful formations but they are not interested primarily in their intrinsic geophysical attributes. They will expect to have fun in a playful way, like the Immortals and enjoy themselves as they journey through the *dong*, the passageway to 'Heaven'. They will play often highly interactive games, with guides asking the visitors to guess the identity of a particular feature, or feel how smooth a karst flow is. For example, in the San Yu Cave (Three Travellers Cave) in the region of the Three Gorges of the Yangtze River, a stalactite is called the 'Sky Bell singing to Heaven' because when the guide invites visitors to strike it with an iron rod a clear sweet ringing sound is heard. In the same cave, another formation is called the 'Ground Drum calling people to the Opera' because when visitors are invited to stamp on it, drum-beats echo around the cavern (Li, 2006: 260). In Tenglong Cave, Hubei Province (at more than

52 km long, the largest in China), there is a central performance cavern with souvenir and refreshments stalls and an hour-long cultural song-and-dance concert that includes a laser sound-and light show depicting *inter alia* scenes from 'Journey to the West' projected onto the rock wall. Many caves feature such cultural performances, particularly if one of China's 55 Minorities peoples inhabit the region. Tea-making ceremonies by traditionally dressed courtiers are also common.

Many caves are adorned with calligraphy such as San Yu Cave which is famous because two trios of eminent poets, the first three from the Tang Dynasty who discovered the cave in 819 AD, and the second trio from the Northern Song Dynasty 137 years later, were so impressed by its beauty that all six inscribed poems on the cave walls. Calligraphy captures numerous elements of Chinese 'high culture' – classical literature and poetry, famous literati, epic events, artistic form, craftsmanship and the sacred symbolism of the origins of the Chinese script as a gift from the gods (Li, 2006).

In caves all over China, visitors are invited to touch and feel and stroke formations for happiness, for good fortune, for longevity, to receive blessings (when the formation represents a god or goddess) – actions designed to provide satisfaction for humans in the Chinese context of man and nature in harmony (i.e. Nature giving humans what they need). Western concerns about environmental degradation are mostly absent. Caves with a pool or underground river often have booths with one or more very old tortoises which symbolize wisdom and longevity in both Daoist and Buddhist lore, and visitors will pay a fee to stroke their back for good fortune.

There will usually be other things to 'do' as well. For example, in one of the five caverns inside Yaolin Cave (Hangzhou), there is an elaborate man-made structure with three bridges (symbolic of longevity, prosperity, good health) constructed over an artificial pool leading to a replica of the Entrance Gate of the Forbidden Palace in Beijing, set among two giant natural columns (stalactites and stalagmites that have joined together). Between the two columns is a natural flowstone that resembles the imperial throne and, for a fee, tourists dress in the robes of the emperor and/or empress, and with the entire scene floodlit, choose which bridge to cross to have their photograph taken seated on the imperial throne. In another of Yaolin's caverns, three artificial dragons each about 5 m long will be animated in a blaze of lights and roaring sounds and for 10 Yuan, tourists can take a photo of them. For 20 Yuan, they can be in the photo themselves! In Bawang Cave, Hubei, there is a formation called 'The Dragon Holds its Head High' and visitors are invited to climb inside its hollow 'skull' for a photo. Western visitors might see it as 'Disneyfication' but for Chinese, these images represent a deep, rich and textured symbolism, a profound expression of an ancient culture (Li, 2006). The very appellation, 'Dragon', encapsulates the culture of China and its people since this mythical creature has been the symbol of the country for several thousand years, embedded in many aspects of its society. Ancient emperors were regarded as 'sons of dragons' and the Chinese people are thus emblematically descendants of dragons. Dragons are credited with being generous and wise, powerful and benevolent and able to summon rain for good harvests, living alongside mankind providing protection and guidance. Thus, unlike their fire-breathing counterparts in Western folklore, Chinese dragons were benevolent creatures with divine origins, despite their fearsome appearance. Situating dragons in caves is thus seen by Chinese as a responsible albeit playful gesture of respect to their ancient culture rather than a meaningless parody (Tan, 2013).

In short, very few caves in China fit comfortably into any of Gillieson's six categories for determining management. Environmental conservation is not the guiding principle; artificial and additional constructions are commonplace and expected in order to 'improve on nature', and entertainment prevails; educative interpretation in a scientific framework is generally absent.

Crown Cave

Crown Cave was one of more than 30 caves that Dr Li researched when undertaking her doctoral studies in China between 1998–2004 and was first visited by both authors in 1999 (Li, 2006) (see Figure 3.1).

Crown Cave is so called because of its resemblance to the traditional crown of China's ancient emperors

Figure 3.1 Crown Cave, China

Another visit was made in 2006, again in 2011, and then with teams of researchers in 2012, 2013 and 2014, and the site was monitored annually from 2015–2017. Each year, five teams, consisting of senior government officials from Southeast Asian countries, with at least one Chinese member to act as interpreter, interviewed five sets of stakeholders, viz: (a) senior staff of the management company; (b) operators and owners of souvenir stalls; (c) village restaurants; (d) village hotels; and (e) horse cart transport and river rafts/boats available for hire. These interviews were designed by the authors as experiential learning fieldtrips (Kolb, 1984) for participants taking place in the annual Asian Development Bank-funded Workshop on 'Tourism Management' hosted by the Tourism University of Guilin, variations of which were held annually from 2011 to 2017. This analysis thus has a longitudinal element with observations spanning almost 20 years.

Crown Cave, on the banks of the Lijiang River near the Minorities Hui village of Caiping, is a very famous cave that has been recorded in Chinese literature for more than 2000 years. It exemplifies many of the cultural aspects of caves in China, but its interpretation is unique because of the main aspect of traditional culture that it draws upon in ways that are different from virtually all other caves in China. In 1995, it was leased for 50 years to a Taiwanese company in a joint venture with the Government-owned Guilin Tourist Company (60–40 shareholding respectively). Initial interviews with Management in 1999 revealed that because the cave's formations were less impressive than many others (souvenir hunters over centuries had denuded it of many stalactites and stalagmites), its development was predicated on the five main elements of the Chinese concept (*Wu-xing*) of the physical universe – wood, earth, metal, water, and fire – to provide different 'journeys' as fundamental to the experience. The 'touristization' of the cave was designed to capture four of the five elements (not fire), together with a sixth element, air, through the innovations shown in Table 3.2 and Figure 3.2.

Despite its relative lack of impressive formations, Crown Cave's interpretation wherever possible follows traditional themes such as those associated with 'Journey to the West'. Thus the monkey king features in several places, and there are other opportunities to touch formations and climb into them, such as putting one's head into the 'Jaws of the Dragon rock'. At another formation described as Lord Buddha, the tour guides encourage their groups to stroke Buddha's rotund tummy for good luck! Crown Cave also has leased concessions at the best formations for photographic opportunities and many artificial decorations such as plastic flowers, shrubs, vines and rainbow-coloured lights grace the formations to enhance the photos ('man improving on nature'). There are numerous shopping opportunities inside the cave with stalls selling souvenirs, refreshments, calligraphy, cave exploration equipment, and so forth. Crown Cave also has a local historical point of interest – traditional rice

Table 3.2 The elements of the physical universe in Crown Cave (Li, 2006)

1.	Wood	A wooden pavilion and Daoist temple were constructed outside the entrance and a wooden carving of the Goddess of Mercy is located at the exit point. They take visitors on a spiritual journey.
2.	Earth	To take people into the cave, a monorail has been constructed over the rice fields for 1.2 km (the first physical journey).
3.	Metal	The monorail is made of steel and tourists drive themselves in two-person electric cars made with a steel frame, into the centre of the cave through an artificially constructed tunnel. Once inside the cave, there is a small electric train that takes tourists between different caverns, and it also includes the two elements of earth and metal (the second physical journey).
4.	Water	A small natural flow of water through the cave disappeared in the dry season so pumps were installed to increase the flow; an artificial waterfall 10 m high and 25 m wide was constructed and an existing 500-m-long tunnel was waterproofed and enlarged to make a waterway (an artificial underground 'river') along which visitors are taken in boats (the third physical journey).
5.	Air	To exit the cave at the top of the peak, a glass-fronted seven-storey tall elevator was installed inside the central cavern. This is the fourth journey. It is spiritual as well physical because of its symbolism for Chinese – in ancient mythology, the founding father of the Yellow Race, Emperor Huang-di Yi, left earth on the back of a dragon to ascend into heaven, and visitors to the Cave emulate the emperor by 'riding' this artificial 'dragon' to the summit of the peak. The element of air is also captured through interpretation of a large natural formation in the cave entitled 'Eagle extending its wings'.

wine was once brewed and kept cool in large stone jars deep inside the cave. Wine-making has been revived and is now an additional souvenir item for sale from the original 'cellar' inside the cave.

Many Chinese visitors understand the symbolism behind the 'journeys' provided for the Crown Cave experience, and because they also accept and understand the 2000-year old tradition of improving on Nature, so the monorail, artificial tunnel/waterway and waterfall, train inside the cave, and elevator 'dragon' are appreciated and add to the satisfaction of many Chinese tourists (see Figure 3.2). While it is an example of Gillieson's sixth category of caves open for tourism, it follows none of the western principles of management and provides a clear difference between the biocentric and anthropocentric philosophies as applied to cave tourism in western countries and China.

'Mainstreaming' for Poverty Alleviation: Responsible Rural Tourism Through CBtT (Communities Benefitting through Tourism)

For the past decade, the two authors have been involved with a diverse set of agencies, including multilateral and bilateral aid agencies, non-governmental organizations and research centres such as the Asian

Figure 3.2 'Journeys' by earth, metal, water and air

Development Bank, the World Bank, the UN International Trade Centre, the Commonwealth Secretariat and the UK Department for International Development, to improve the terms of engagement of impoverished communities with tourism to increase opportunities to lift them out of poverty. We have re-evaluated our approach to pro-poor tourism to link communities into mainstream tourism because it can provide many more opportunities for poor people more so than small community-based tourism (CBT) projects. A more market-led demand-side approach with a range of different partnerships (instead of just focusing on community ownership) has been adopted. These development assistance organizations have concluded that if a broader socioeconomic impact is to be achieved, then it is essential to try and 'capture' or harness substantial existing tourist flows where economies of scale offer the opportunity to engage/employ many more impoverished people than otherwise. This can be achieved by 'mainstreaming', that is linking impoverished communities with large-scale and/or high-yield tourism flows. Supply chain and value chain analysis provide tools to identify opportunities for such communities to benefit from existing tourism flows.

Community ownership is not viewed as fundamental, with large-scale tourism creating many opportunities, some direct some indirect, for income generation in other ways. We have coined the term CBtT, 'Communities Benefitting through Tourism', to describe this phenomenon (Sofield, 2008). Li (2017) has compiled a set of case studies on mainstreaming CBtT ventures that have successfully lifted communities out of poverty in Myanmar, Indonesia, Vietnam and the Philippines. For example, Compostela Valley Province, Mindanao, Philippines has well-developed beach resort tourism that attracts large numbers of domestic tourists as well as international visitors. The local fishing village of Tagnanan near the main resort area was marginalized, so tourism was used in two ways to engage the community in productive income-earning activities. The first was to train fishermen as tour guides, a viable alternative because of overfishing that had reduced incomes. The second was to train fishermen's wives to become involved in horticulture, growing and selling plants and flowers from stalls set up at the entrances to all the different resorts. Both of these activities are a form of CBtT and many households in the community have been lifted out of poverty because of the opportunity to tap into the major tourist flows to the nearby resorts (Li, 2017: 33).

The development of Crown Cave for tourism provides an example of a public/private sector partnership and mainstreaming that has eliminated poverty in the nearby village of Caoping, inhabited by the Hui Minority people. As noted, the cave was developed as a joint venture between the Guilin Tourist Company (government-owned) and a private sector investor from Taiwan. The Provincial Government provided the basic infrastructure – a sealed road for 30 km from Guilin to Caoping village, together with electricity. The private entrepreneur then built all the amenities to visit the cave–the monorail to access the cave, a pier on the river for access from the Lijiang River, paths, tunnels, stairs, toilets, etc., a seven-storey high glass-fronted elevator inside the cave, and the artificial waterfall and underground river with water diverted from an irrigation dam originally built by the government for farming. Total development expenditure from 1995 to 2014 was more than RMB300 million (almost USD50 million). The mass visitation to Guilin – about 12 million in 2010, 27 million in 2017 – was targeted to provide large-scale visitation to the cave to justify the high capital cost of development.

Inside the cave, 20 souvenir stalls and refreshment kiosks, and two photographic opportunity kiosks were set up by a sub-contracted company which leases them to local villagers for RMB5000 per month (these numbers were being expanded on our most recent visit in 2017). In Caiping village, in addition to the Crown Cave company management office, a small conference centre and the monorail station, a souvenir shopping mall was also constructed and 30 shop spaces are leased to the local people. These various activities and facilities employ more than 150 local people, where prior to the development of Crown Cave there were no local paid jobs other than ten government administrative posts, three school teachers and a small two-person clinic.

As numbers of visitors to the cave increased over the years (to about 1 million visitors p.a.), the local people tapped into the flow of tourists to establish many other businesses. Our investigations identified another 30 souvenir stalls and 10 refreshment stalls dotted around the village, also five restaurants and four small hotels owned by village families. There are 40–50 horse-drawn carts for alternative transport along the river bank from the village to the cave, and more than 50 small boats and river rafts for river tours (less than ten families continue to engage in fishing as a regular activity). Each household is engaged in multiple businesses – 80% still farm (1–2 household members) with honey as a new product that has proved popular with tourists. Most of the food for the restaurants is produced locally thus creating a strong linkage between the local farming and fishing activities around the village and the tourism sector. There is a small local market in the village where tourists purchase fresh fish, fruits and vegetables on a daily basis but we have no figures for income generated by this activity. Some 90 households have one or more members of the family employed by the Caves company (note that as a Minorities people, the Hui were not bound by the one-child policy, hence 2–3 children per family, occasionally more, is common).

The figures set out in Table 3.3 with the exception of the wages for staff employed by the Crown Cave Company, should be regarded as indicative rather than definitive: local village people are in general reluctant to divulge exact levels of income.

However, repeated interviews with stakeholders over a three-year period using different techniques (such as asking a restaurateur how many lunches they sold in peak period per day and in low season, and the average cost per meal, an estimate of annual income could be calculated using extrapolation) are considered to have provided sufficient detail (allowing for an error of 10%) to demonstrate convincingly that all households in

Table 3.3 Tourism-generated income for Caiping Village 2014

Type of business	No. owned by villagers	Earnings for each business – RMB p.a.	USD
Souvenir stalls	30+	36,000–100,000	5100–15,700
Refreshment kiosks	10	40,000–55,000	5700–8000
Small hotel/homestay	4	65,000–122,000	9300–17,500
Horse cart transport	40–50	30,000–35,000	4300–5000
Boat, raft river tours	50	40,000–50,000	5700–7100
Village restaurants	5	100,000–140,000	15,700–20,000
Villagers employed by Crown Cave Co	150		
Average wages of each villager p.a.		30,750	4410
Total wages of villagers by CC Co. p.a.		4,612,500	658,928
Estimated average household income 2014		100,000+	15,700+

Caiping have been lifted out of poverty. Prior to cave tourism, government statistics indicated that the per household income was less than RMB 9000 p.a. (USD1200) as against more than RMB100,000 (USD15,700) in 2014.

Our surveys of village businesses demonstrated that tourism 'mainstreaming' has transformed this previously poor farming and fishing village into relatively wealthy rural households. We were unable to obtain specific details of the total wages paid by the company to villagers over the period 1995–2014, but even in the early days of operations it exceeded more than RMB1 million and a conservative figure for the 20-year period would amount to about RMB40 million. When locally generated incomes through CBT/CBtT activities are added, that figure is probably doubled. Without mainstreaming, whereby the Crown Cave Co was able to tap into the millions of visitors to Guilin annually and attract 1 million of them to the cave, it is inconceivable that the Caiping community could have generated such an income by themselves through a 'normal' small-scale community owned venture. In this context it is important to note the size of the investment made by the Company in developing the cave and associated amenities and facilities, without which the village households would have had little or none of the opportunities that have been created and which have lifted them out of poverty. Although the tourist activities take place in and around the village, the cave visitation is not CBT since there is no local ownership. Some of the village-owned activities are a form of CBT, but the overall situation (especially through employment by the Crown Cave Co) is CBtT (Li, 2017; Sofield, 2008).

Conclusion

In summarizing the cultural differences between cave tourism in western countries and in China, it is relevant to refer to Kirk (1963) in his review of the problems of geography where he identified a dichotomy between the phenomenal environment – of empirical facts about phenomena; and the behavioural environment – the environment as perceived. These same differences apply to the management and interpretation of caves for tourism in western societies and in China. Western management and interpretation is built around scientific, empirical fact. The emphasis is on education. The Chinese interpretation is perceived, symbolic and emanates from millennia of cultural heritage: the emphasis is on entertainment and we thus have a contrast between the playful versus the educational. In short, it is invalid to use western concepts of environmentally oriented management to Crown Cave's tourism and criticize it as irresponsible.

In Chinese terms it provides a satisfying rural tourism experience that is deeply embedded in a range of cultural values that are simply different from western ideas which lack a similar continuous 3000+ years heritage of cave use and visitation. Crown Cave is described as 'rural ecotourism' in official brochures and its web site, in common with many rural sites all over China that have in fact been strongly engineered and manipulated by humans, with such changes perceived as culturally consistent, culturally acceptable and not disparaging of authenticity. The mechanization of transport to and inside the Crown Cave, the modification of the cave with manmade linking tunnels, artificial waterfall and river, a seven-storey-tall glass-fronted elevator inside the cave and all the many entrepreneurial activities in its caverns, provide a stark contrast with the environmentally sensitive management, operations and scientific interpretive tourism of caves in western countries.

The combination of partnerships and co-existing activities involving community/cultural/ commercial factors that can lead to poverty reduction for communities revealed by our Crown Cave research, force a rethink about aspects of PPT (pro-poor tourism) and the orthodox approach adopted for most CBT ventures which are based on local ownership. The revised concept of CBtT is a valid alternative.

At this juncture it is pertinent to ask whether Crown Cave could have been developed by the community itself? Would the government or some other aid donor have been willing to invest such a large sum of money in a community-based tourism venture?

First, a relatively smaller amount of money might have been provided (say USD0.5 million or RMB3.5 million, equivalent to other CBT projects elsewhere). Thus the answer is in the affirmative: the cave could have been developed with little infrastructure and owned by the Caiping community. But such an investment could not have produced the large incomes that the Caoping village households have been able to earn from a diversity of employment opportunities, support services and small businesses that they now operate because of the involvement of 'big business' in Crown Cave that has generated mass tourism on a scale well beyond 'normal' CBT ventures. Incomes would have been much, much smaller, and in common with most CBT ventures, many household members would not have been earning wages in running the family business. Second, because the villagers lacked technical, business and management skills, and had no understanding of how to market the cave and place it on the tourist map, based on the experience of other CBT projects elsewhere, it would have taken ten years or more to provide training and reach economic viability based on perhaps 5000–10,000 visitors p.a. as a small self-sustaining business without outside assistance. With the involvement of 'big business' however, benefits started flowing to the community on a substantial level from Year One. Third, a community-owned CBT enterprise could not have achieved the scale of poverty alleviation achieved by the large outside investment in developing Crown Cave through the public/private sector partnership.

Is it responsible rural tourism? It is undoubtedly rural, since it is located more than 30 km from the nearest city (Guilin), and while the nearby village of Caiping has inevitably been modernized to a certain extent, it still retains key characteristic of a rural community. In applying the definition of responsible tourism utilized for this volume, i.e. 'Responsible tourism is tourism which minimizes negative social, economic and environmental impacts, generates greater economic benefits for local people and enhances the well-being of host communities' (Goodwin, 2007 based on the Capetown Declaration 2002), it can be argued that Crown Cave meets all of these requirements, with the possible exception of environmental impacts. But even this point can be contested given that the ecology of Crown Cave had been altered over centuries of visitation and that its contemporary development for tourism in Chinese terms is valid and responsible. Its economic benefits for and social well-being of the local community are clearly demonstrated. While most of the literature on empowerment emphasizes community ownership of projects (Honey, 2008; Scheyvens, 1999), in Caiping the level of income generated by cave tourism has removed households from a 'survival' mode of living standard to one where for the first time in their lives, they are free to make decisions never before possible. This includes not only the capacity to purchase consumer goods, appliances and equipment that ease the burden of physical labour and time spent on mundane chores, and in installing electricity and plumbing in their homes for improved health outcomes, but also in sending their children to university – the first generation of Hui to have this opportunity. Discussions with them about their substantially higher living standards do include complaints about the dominance of the Crown Cave Co, but they also acknowledge that the benefits have changed their entire lives and lifted them out of poverty: they have been empowered through an ability to make choices they themselves determine in areas never before conceivable. This case study thus also presents a challenge to the prevailing thinking among many western academics that community ownership is the sole avenue of/for empowerment.

Crown Cave tourism and Caoping village exemplify the advantages of 'mainstreaming' to achieve poverty alleviation on a much larger scale through CBtT than is rarely possible with the more normal small CBT projects. We argue that it constitutes an example of sound, sustainable and responsible rural tourism in China which is rooted in the Asian philosophy of harmony between humans and nature.

Acknowledgements

Grateful thanks are due to:

- The Asian Development Bank which through its Tourism Management Workshops in Guilin, Guangxi Province, China, from 2011 to 2017, facilitated our research in Crown Cave and Caiping village.
- The Guilin Tourism University and its staff who played host to the annual workshops.
- The 75 senior government officials from Cambodia, China, Laos, Myanmar, Thailand and Vietnam who formed our research teams to interview stakeholders each year.
- The management of the Crown Cave Co who willingly made themselves available for a round-table discussion each year.
- Last but by no means least, the villagers of Caiping who accommodated our intrusive enquiries every year.

(All photographs taken by the authors).

References

Chan, Wing-tsit (1969) *A Source Book of Chinese Philosophy*. New York: Colombia University Press.
Gan, Weilin (1988) The duties of scenery and heritage sites. In W.H. Ding, Y.M. Xu and Y.X. Lin (eds) *The Study of Natural Scenery and Heritage Sites* (ch. 3, pp. 61–64). Shanghai: Tongji University Press.
Gillieson, D. (1996) *Caves: Processes, Development and Management*. Oxford: Blackwell Publishers.
Goodwin, H. (2007) What is Responsible Tourism? Accessed 20 June 2018. http://responsibletourismpartnership.org/what-is-responsible-tourism/
Ham, S. (1992) *Environmental Interpretation: A Practical Guide for People with Big Ideas and Small Budgets*. Golden, Colorado: Fulcrum Publishing.
Honey, M. (2008). *Ecotourism and Sustainable Development: Who Owns Paradise?* (2nd edn.). Washington, DC: Island Press.
Hughey, K.F.D. and Ward, J.C. (eds) (2003) Sustainable Management of Natural Assets Used For Tourism in New Zealand: A Classification System, Management Guidelines and Indicators. Tourism Recreation Research and Education Centre (TRREC),

Report No. 55, Lincoln University. Wellington: Department of Conservation and Tourism New Zealand.

Kirk, W. (1963) Problems of geography. *Geography* 48, 357–371.

Kolb, D. (1984) *Experiential Learning: Experience as the Source of Learning and Development*. Englewood Cliffs: Prentice-Hall.

Li, F.M.S. (2006) Chinese common knowledge, tourism and natural landscapes. Gazing on 别有天地 *'Bieyou tian di'* 'An altogether different world'. PhD thesis, Murdoch University, Western Australia.

Li, F.M.S. (2017) Community tourism: Partnerships for poverty alleviation and 'mainstreaming'. *Workshop on Tourism Management*, Asian Development Bank, Guilin, China: 20–27 April 2017.

Scheyvens, R. (1999) Ecotourism and the empowerment of local communities. *Tourism Management* 20 (2), 245–259.

Sofield, T.H.B. (2008) The dolphin discovery trail: Seeking the balance between ecotourism and conservation for poverty alleviation. In J. Bao, H. Xu, T. Sofield, J. Sun and L. Ma (eds) *Tourism and Community Development – Asian Cases* (Ch. 3, pp. 57–80). Madrid: UNWTO.

Sofield, T.H.B. and Li, F.M.S. (1998) China: Tourism development and cultural policies. *Annals of Tourism Research* 25 (2), 362–392

Stitt, R. (1994) US Forest Service: FCRPA Regulations. *Cave Conservationist: The Newsletter of Cave Conservation and Management* 13 (3), 1–15.

Tan, A.S. (2013) The Chinese Dragon, God's Frolicking Dragon and the Cross of Christ. In M.P. Maggay (ed.) *The Gospel in Culture: Contextualization Issues through Asian Eyes* (ch. 12, pp. 132–150). Manila: OMF Literature Inc.

Xu H., Qingming C., Sofield, T.H.B. and Li, F.M.S. (2014) Attaining harmony: Understanding the relationship between ecotourism and protected areas in China. *Journal of Sustainable Tourism* 22 (8), 1131–1150.

Theme 2
An Essentially Asian Form of Rural Tourism

4 Responsible Rural Tourism in Japan's Tea Villages

Lee Jolliffe and Moe Nakashima

Introduction

In Japan, tea is produced mainly in small villages and the surrounding rural areas. These settlements with their rolling tea landscapes, traditional cultures and long traditions of the production and consumption of green tea provide authentic settings for small-scale tourism. Activities within such tourism will ideally be organized by locals with the benefits going directly to them on an ongoing basis. Stakeholders engaged in this type of rural tourism development may include local associations, tea farm owners, operators of small cafés and shops, village residents and visitors.

This chapter considers the emergence of tea-related tourism as a form of rural tourism in a number of Japan's tea villages. It draws on secondary resources on village-scale tea tourism in Japan, and on direct experiences of the authors with one of the tea villages. This involvement included participant observation during a tea tour and a three-month internship (2017) as well as an intensive field visit (2018) where we conducted focused interviews with community stakeholders. The case study methodology using multi-methods has been identified as being suitable for tourism research (Beeton, 2005). There is value in using multiple cases that can be examined for within-case and cross-case analysis (Brotherton, 1999).

In terms of methods, the chapter uses document analysis, participant observation and focused interviews with stakeholders for the central case, while additional cases based on secondary sources are presented for comparison and analysis. Ethics approval for this research was received from the University of New Brunswick. A limitation of the work is that it investigates the development of responsible rural tourism from the local community level and does not consider the involvement of destination management organizations. The set of case studies includes one village each from three of Japan's significant tea-producing prefectures, Higashi

Sonogi (Nagasaki), Wazuka (Kyoto) and Osawa (Shizuoka). The central case is that of Wazuka. A map shows the major areas of tea production in Japan as well as the regions associated with our case study sites (Figure 4.1).

Background Information

Rural tourism

Rural tourism involves the concept of rurality that includes three elements (Lane, 1994). These are population density and settlement size; land use and economy with the dominance of agriculture and forestry; and traditional social structures with accompanying issues of community identity and heritage. In rural areas undergoing restructuring, tourism is often seen as a solution to rural economics and development, but if not properly implemented it can threaten the social composition and rural environment (Roberts *et al.*, 2017). Rural areas tend to harbor older ways of life that are sought out by visitors as an antidote to urban living (Urry, 1988). Tourism in rural locations may be connected to local agricultural

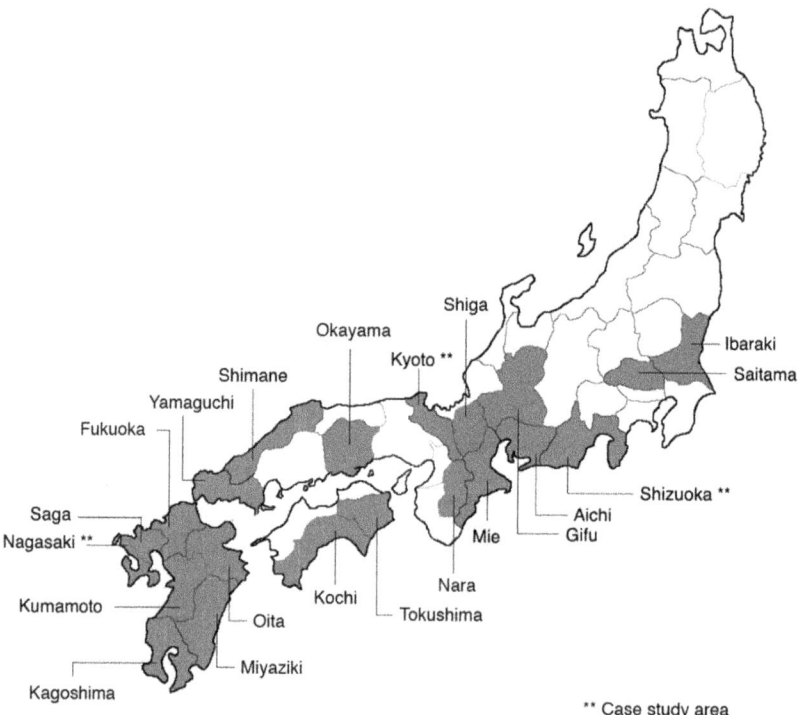

Figure 4.1 Tea-producing areas of Japan with case study areas
Source: Logan Jolliffe.

Table 4.1 Characteristics of integrated rural tourism

1	Ethos of promoting multidimensional sustainability
2	Empowerment of local people
3	Endogenous ownership and resource use
4	Complimentary to other economic sectors and activities
5	Appropriate scale of development
6	Networking among stakeholders
7	Embeddedness in local systems

Source: Adapted from Cawley and Gilmor (2008: 319).

products, as in Japan where farm brands can lead to complimentary economic effects from tourism activity (Ohe & Kurihara, 2013).

Ideally, rural tourism development takes into account various rural resources (cultural, social, environmental, economic), their use and the role of related stakeholders, in a planning approach that Cawley and Gillmor (2008) call 'integrated' rural tourism (Table 4.1). This has an emphasis on a bottom-up approach involving local stakeholders utilizing resources incorporating sustainable principles to make optimal use of assets while protecting and enhancing them (Aylward & Kelliher, 2009).

This 'integrated' approach to responsible rural tourism provides a foundation for examining tea-related tourism as a type of rural tourism occurring in Japan's tea villages. In the light of large-scale urban migration in Japan, tourism has become more important to the depopulated rural areas (Knight, 1996). For such areas, tourism has held the promise of economic revival and rural tourism is seen as a way of revitalizing communities (Arahi, 1998). Tourism in rural Japan includes small-scale local activities resulting from sustainable partnerships, as reflected by the development of green tourism in rural areas (Arahi, 1998; Hashimoto & Telfer, 2010).

Tea tourism

Tea tourism, that is, tourism motivated by an interest in the history and culture of tea, may take place in rural agricultural areas where the tea is grown and processed (Jolliffe, 2007). As such it can be considered to be a form of rural tourism. A study of tea tourism in rural Sri Lanka identified potential for developing such tourism through community collaboration while noting challenges that include limited infrastructure, lack of capital, seasonality and absence of relevant government policies (Fernando et al., 2017). In Sri Lanka tea-related tourism in areas of production has been identified as a tool for marketing rural destinations (Fernando et al., 2016; Jolliffe & Aslam, 2009). Tea tourism in such areas can potentially be a form of responsible rural tourism. For tea tourism in these areas to succeed, stakeholder collaboration is required, as reflected by an investigation in rural China (Cheng et al., 2012). Another study in China has shown that tea tourism has the potential to benefit tea farmers and tea merchants (Chee-Beng & Yuling, 2010).

In Japan, the development of green forms of tourism has encouraged tourist involvement in farm activities such as tea-leaf picking (Knight, 1996). Green tourism in Japan has contributed to increased interactions between urban and rural residents that are characteristic of rural tourism (Bixia & Zhenmian, 2013). The promotion of tourism to villages with features that include forested mountains, meandering rivers and old houses is encompassed by the term *Furusato*. This means hometown and connotes a desirable lifestyle aesthetic with rural vistas for visitors to enjoy (Robertson, 1988). The concept has a connection to the rural places away from cities. It emerged in part due to the revitalization of rural villages in Japan (Knight, 1994). This type of nostalgic tourism in such rural locations can form an extension of farming activities and provide additional sources of income for farmers while respecting the environment and traditional cultures as reflected by green tourism in rural Kyushu, Japan (Hashimoto & Telfer, 2010).

Tea Tourism Context in Japan

Tea has been produced in rural Japan since as early as the 9th century when it was consumed by the religious classes who had gone to China to learn about it. It was further popularized by the gentry in the 12th century and encouraged by the 2010 publication of the *Way of Tea* by Aaron Fisher. Today, most of the tea produced is green tea and largely consumed in Japan, with about 3% being exported (Japan Tea Export Council, n.d.). This provides an incentive for international visitors interested in Japanese tea to visit and experience the tea at the source. The culture of Japanese tea is well known through the tea ceremony, an evocative symbol and a source of nationalism (Pitelka, 2013). The powdered tea (Matcha) used in the tea ceremony has been popularized through its use in beverages by international chains such as Starbucks.

The growing and processing season for tea is typically from May to October, when activities related to tourism typically occur, as visitors are interested in seeing and experiencing the tea during this time. Tea-related tourism in these areas is thus subject to seasonality. Touristic activities related to tea during the growing season may include independent (walking, hiking or cycling) or guided visits to tea fields, tea tastings, participation in tea festivals, homestays and other tea-related special events (Jolliffe, 2007). These activities are typically small scale as there is limited infrastructure (transport, accommodation and food service) in the rural tea producing areas for hosting large numbers of tourists. There is some limited potential for off-season activities in terms of tea tasting and related instruction.

Case Study 1 – Higashi Sonogi (Nagasaki)

The tea-farming community of Higashi Sonogi is located in Nagasaki Prefecture, Kyushu, in the south of Japan. Nagasaki is one of the smaller tea-growing areas in Japan. The town itself has an area of 74.29 square kilometres and a population of 8175. It is accessible by rail with regional and local trains (Japan Rail Kyushu) and has a historic railway station. There is also a local bus service. The town has several cafés where the local teas can be experienced. There are approximately 145 tea farmers in the town. The town produces 60% of the tea in Nagasaki Prefecture. The history of tea farming at this location dates back over 500 years to the 17th century Edo period when the cultivation of the tea fields began.

A group of area tea farmers created what they describe as a new form of eco-tourism, called 'Green Tea Tourism' to promote exchange between rural and urban communities (Higashi Songi Green Tea Tourism Association, n.d.) After initial efforts in 2014, it was founded in 2015 by tea farmers and local officials. The five principle activities of the association reflect the traditional Japanese tea ceremony referred to as Chadō (茶道) the *Way of Tea*, a leading principle of hospitality deeply rooted at Higashi Sonogi (Table 4.2). By 2016, the community had introduced home-stays with tea activities in the village. As part of the initiative, three tea farmers offer homestay accommodation for visitors in a traditional Japanese house with related activities that may include visiting tea fields and learning how tea is processed and prepared. Tea tourism here is often part of a family business, for example tea farmer Ooyama Yoshiaki is a fourth-generation owner of the Ooyama tea farm and a member of the association, his mother is a certified tea instructor who runs a tea shop and has operated a guest house since 2016 (Ikkyu Tea, n.d.).

Table 4.2 Objectives of Higashi Sonogi Green Tea Association

Principle	
1	Share Sonogi Tea with the world, contribute to increasing the number of Japanese green tea drinkers.
2	The scenery of the tea fields in Higashi Sonogi is a treasure we can share with the world.
3	We celebrate the ties that bring the people of Higashi Sonogi together.
4	We will share the appeal of Higashi Sonogi's cuisine and local foods.
5	We will create a community where everybody can feel the comfort and warmth known locally as 'Unbantsukura'.

Source: http://greentea-homestay.com/#association

The type of experience here is reflected by the comment of one visitor:

> 'It's like one big green carpet!! This is the kind of beautiful view you can only get in Higashi Sonogi, the land of tea. Higashi Sonogi-cho is where around 60% of all the tea in Nagasaki Prefecture is produced!' (Tomocchi, 2017)

The tea tourism programme here fits the objectives of integrated rural tourism noted earlier in the chapter. The goals are closely related to those of ecotourism and green tourism that have been promoted by the Japanese government as a means of rural regeneration (Arahi, 1998). At this rural tea-producing location, the emphasis for tea-related tourism has been on homestays and the tea experience, where visitors can learn tea production and culture. That this location, while rural, is accessible by rail is surely an advantage for further development of tea tourism as a form of responsible rural tourism with a strong level of community involvement.

Case Study 2 – Wazuka (Kyoto, Uji)

The small tea farming village of Wazuka located in Kyoto Prefecture is in the Uji tea region. This is Japan's oldest tea-growing areas and is well known for tea production. Wazuka has been designated as 'One of the most beautiful villages' of Japan under a programme founded in 2005 to identify villages that wanted to protect their natural and social capital as intangible resources for the future. The total area of the village is less than 65 square kilometres.

With a declining population of less than 4000 and over 300 tea farming families, small-scale tourism has been introduced as a means to encourage the appreciation of Japanese tea and to nurture the interest in tea farming. International short-term work camps established since 2001 have encouraged young people to become tea farmers, rejuvenating abandoned tea farms (Service Civil International, n.d.).

Kyoto Obobu Tea Farms (Obobu)(n.d.) has been established as a social enterprise, with the mission of bringing Japanese tea to the world. This entity now hosts one-day tea tours, provides tea focused three-month internships, hosts tea picking and processing events a few times a year and conducts a two-week master class a few times a year. The Wazukcha Café acts as an informal welcome centre providing information, local teas, tea-related food, tea ware and tea-dyed product souvenir sales, and bicycle rentals, etc. Several other cafés feature local tea and cuisine, serving food and with tea for sale. One, D-Matcha, has a focus on matcha and Wazuka's green tea history. Throughout the community, there are interpretive signs related to tea production. A free Wazuka Gourmet Map & Guide highlights tea landscapes and cuisine in Wazuka.

Tea production in Japan has a rich heritage, and the over 800-year history in Wazuka is portrayed by Kyoto Tourism as the 'cradle of Japanese tea' (Zavadckyte, 2013). A significant percentage of Japan's Matcha (24%) is produced here. Activities promoted for tourism include walking, hiking and biking with the backdrop of the tea fields, tea tasting and experiencing tea culture and related cuisine, at cafés and through organized tours as well as shopping for souvenirs related to tea and local goods.

A weekend Teatopia Festival established by the Wazuka community in 2012 and now held annually in early November features a number of small suppliers of tea-related experiences. Here visitors can try different types of tea and experience tea-related cuisine. According to one stakeholder interviewed, attendance has more than doubled since establishment, from 5000 in 2012 to 11,000 in 2017. With limited accommodation capacity, the challenge is to spread visitation out over the season, rather than receiving so many at one time. An issue with the continued development is addressing seasonality as the destination is more appealing during the tea production season.

One informant observed that tea tourism in Wazuka included a number of activities. 'Mainly (the) tea tour, drinking tea, going to (the) tea field, picking, factory tour, observing processing (depending on the season)' (Tea Farmer).

In addition to the challenge of seasonality, major issues facing the development of tea tourism here are limited transportation access and accommodation. Wazuka is served by a public bus from Kamo, on the Japan Rail line, with connections from Kyoto, Osaka and Nara. There are several accommodation facilities operated including Blodge Lodge, an AirBnb property and Kyoto Wuzukaso, a *ryokan* (traditional Japanese inn) with four rooms featuring traditional food and tea baths.

Local stakeholders in tea tourism in Wazuka include tea farmers and café and shop owners. Some tea farmers are very involved in tea tourism, such as the co-founder of Obobu who spends much of his time instructing tea tours, working with interns, hosting their annual tea master class and their tea picking and rolling events held to celebrate the different seasons of the tea harvest. He travels annually to Europe to provide education on Japanese tea, strengthening the network of those interested in tea from Japan and Wazuka. The interns at Obobu could be considered to be stakeholders in tea tourism while they are at the village and beyond their stay. Interns assist with the tea tour programme and take on special projects such as blogging about the tea on Obobu's social agricultural enterprise web site, creating informative videos and generally promoting the tea tourism experience.

Stakeholders interviewed in Wazuka included providers of tea-related services (tea tour and course operator, sweet shop owner, tea shop owner, craft shop owner and tea instructor). Several of these stakeholders are also tea farmers. Three of the interviewees are part of third generation traditional businesses with a history of from 60 to 100 years and the other two are newcomers to the area. Questions were asked about their business in terms of establishment and what it provides, and their role in tea-related tourism here. Thoughts about the future directions of tea tourism in the region were elicited. Through these interviews, insights were gained into the development of tea tourism in Wazuka as a responsible form of rural tourism as well as the issues it currently faces.

Many interviewees reported a slow and gradual development of tea-related tourism in the village, attracting domestic and international tourists. 'Wazuka being close to city and attracting international tourists for tea made it even more attractive for domestic tourists, film crews visit the area frequently to market the place' (Tea Shop Owner). 'Before the festival was established in 2012, there was little (tea-related) tourism in Wazuka. It (the festival) brought together several other tea tourism providers who usually offered their services separately (so) for the tea festival all providers came together to welcome visitors at that time' (Tea Tour and Course Operator).

The interviewees want to share their knowledge of the local tea, so they welcome visitors, both locals and tourists. 'There was a demand for (a) place to enjoy tea as there is a lot of tea in Wazuka but nowhere to enjoy it. I wanted to create a place where people could enjoy good tea and are exposed to a wider selection of it' (Tea Shop Owner).

Stakeholders interviewed are open to the growth of tea-related tourism in Wazuka. They do not, for the most part, anticipate any negative effects from visitation to the tea village. There is some concern about the large number of people visiting during the annual tea festival and thoughts have been expressed as to how to disperse this visitation to other times of the year: 'The (tea) festival is a strong and weak point as it is difficult to welcome so many people at once' (Tea Tour and Course Operator).

'Tourism in Wazuka is still not as developed. It is too early to worry about it (in terms of development)' (Craft Shop Owner). 'The people who have been visiting Wazuka have been polite and respectful. Therefore, I am welcoming the fact that Wazuka could become a more popular tourist destination' (Tea Shop Owner). 'Happy there is an increase in tourism' (Tea Instructor). 'Would welcome an increased number of tourists in the area' (Sweet Shop Owner).

The declining population was noted as a challenge to the continued development of tea-related tourism here. This reflects the depopulation of rural areas of Japan noted earlier in the chapter. 'Another issue is that the population of Wazuka is ageing and declining, from about 6000 of 20 years ago to about 4000 today. This very rural area needs newcomers' (Tea Tour and Course Operator).

Stakeholders acknowledged seasonality as an issue, with visitation concentrated around the tea season and the tea festival. 'Seasonality is an issue, so we need to find a way to make it (visitation) more year-round' (Tea Tour and Course Operator).

These stakeholder comments reflect an embeddedness of tea-related tourism within the community with providers welcoming visitors as well as in several cases new entrants into the community as tea farmers and artisans. In this rural village area with a declining population, this is a good sign for the integrated development of responsible rural tourism here.

Case Study 3 – Osawa (Shizuoka)

Shizuoka is a leading tea-growing area in Japan. Within this region in the small tea village of Osawa, surrounded by mountains and tea farms, twice a month on a Sunday, community members open some heritage houses (Nao, 2017). They offer tea and snacks in an event called the Osawa Engawa Café. The establishment of this programme stems from the villager's desire to promote their 'Shizuoka Motoyama tea'. A small fee is charged at each house visited for the local tea and accompanying snack food.

All of the 23 households here participate in the café event from time to time with about nine of them being part of each event. The community is directly participating in the promotion of tea from their area, creating social capital through the interaction of locals with visitors over tea and snacks. The pairing of traditional snacks with tea provides an added element, as tea in Japan is often connected to cuisine. A Japanese travel magazine describes the authentic experiences, characteristic of rural tourism generated by this experience:

> It is impossible to feel the heart of a Japanese farming village culture unless you have friends or acquaintances living there. However, the Osawa Engawa café allows you to feel and experience the people, cultures, and food in Osawa village to your heart's content. (Nao, 2017)

This case demonstrates how tea-related tourism can be introduced on a small scale in harmony with integrated rural tourism principles. The sharing of tea and knowledge about it is a common characteristic of small-scale tea tourism in Japan and in this small village there is a high level of community participation and a desire to share their tea culture with visitors. However, access is limited as there is no public transportation. The offering is available twice a month so this provides restricted access to an authentic tea experience in a tea-growing village.

Discussion

The three rural tea farming case study locations are compared (Table 4.3) within the framework of integrated rural tourism (Table 4.1).

Table 4.3 Comparison of case study locations as integrated rural tourism

Factor	Case 1 – Higashi Sonogi	Case 2 – Wazuka	Case 3 – Osawa
Multidimensional sustainability	Tourism respects local cultural, natural environment, with some economic benefits.	Tourism compliments local tea production, respects local culture and environment, has economic benefits.	Complimenting local environment and society tea tourism here is sustainable.
Empowering local people	Locals empowered through Tea Tourism Association.	Through annual event, local tea farmers and providers share tea knowledge.	Tea café of local households sharing tea and snacks with visitors.
Endogenous ownership/ resource use	Local ownership and operation respecting resource of tea fields and landscapes.	Local enterprises involved with appropriate resource use.	Locally owned use of community cultural resources.
Compliments other economic sectors/ activities	Green tea tourism compliments tea farming, tea farmers host homestays.	Tourism compliments work of tea farmers, café owners and artisans.	Café event compliments area tea farming, serves to promote local tea.
Appropriate development scale	Three of six local tea farmers (Tea Tourism Association founders) provide homestay.	Small scale with a gradual increase in development over time.	Very small scale, local households participating.
Networking among stakeholders	Through an association. Tea hospitality achieves social goals.	Hosting visitors at events involving numerous stakeholders.	Through tea café, between the tea farming households and visitors.
Embedded in local systems	Part of tea production system and industry regeneration, young tea farmers from tea families participating.	Tea industry renewal and regeneration, internships and work placements encourage new tea farmers.	Tea café with local community hosting event in their homes.

All are part of the regeneration of the rural areas of Japan in efforts to counter declining populations and offer a respite for urban dwellers. The objectives of this responsible form of rural tourism at these places primarily relate to sharing the knowledge and experience of Japanese green tea culture with locals and visitors as well providing traditional hospitality over tea. All locations have social goals of preserving the environment and local cultures, consistent with the multidimensional sustainability principle of integrated rural tourism. This is evident in the case of Wazuka, in terms of their membership in 'The Most Beautiful Villages in Japan' programme, whose objectives are similar to those of responsible forms of rural tourism.

The products developed for tea-related tourism at the study locations differ in terms of form, representing a range of experiences including homestay, tours, courses, instruction and festivals. These products are locally produced and have common goals of appropriate resource use while contributing to the rural regeneration of communities. In all cases, forms of responsible tourism compliment agricultural tea production, which is the main industry. Such endeavours value local landscapes and culture in keeping with the goals of integrated rural tourism. Overall the cases represent a deep appreciation of the tea heritage, landscape and traditions and a responsible approach to sustainably operating tea-related tourism on a small rural scale.

An appropriate and responsible scale of development is reflected at all locations. The rural scale is evident, with Higashi Sonogi and Wazuka representing large land areas with low populations while Osawa is a sparsely populated rural village. The cases reveal small-scale enterprises providing tea experiences demonstrating tea farming community stakeholders joining together, through partnerships (for example: association, event, café) to share knowledge of their tea. The cases reflect the Japanese concept of *Furusato* discussed earlier in the chapter, in that through their hometown tea culture they portray a desirable lifestyle aesthetic in a rural setting for visitors to appreciate.

Significant connections to traditional heritage and societies are evident, with tea farming at all locations dating, back many centuries. The cases reveal instances of tea tourism activities strongly connected to local families, either those who have traditionally inhabited the area (at all case study locations) or including newcomers to tea farming and related endeavours (as at Wazuka). With all rural areas of Japan facing a declining population, Wazuka has encouraged the settlement of new tea farmers through work programmes and the internships offered by the community and Obobu. Traditional farming families and their descendants are leading tea tourism in Higashi Sonogi while local families settled within the last decade (newcomers) and small-scale tourism operators are championing tea tourism in Wazuka.

For the case studies considered, responsible rural tourism development results from networking between the tea farmers, households and related providers of tea-related services. This meets the objectives of integrated rural tourism in a responsive manner. Local farmers and citizens are directly involved through formal associations (as at Higashi Sonogi) or informal organization of events (as at Wazuka and Osawa), contributing to local sustainable rural tourism development. Tea-related responsible tourism lies at the heart of these tea-farming communities. As part of the system of tea production and rural regeneration, it adds value in terms of the promotion of the tea and education about it, as well as offering a setting whereby hosts can share the traditional tea culture with visitors in a natural environment. This links back to the concept of green tourism described earlier in the chapter. It is likely that most visitors want to share in the traditional culture surrounding tea while experiencing nature, rather than have an immersive tea tourism experience. However, as this aspect was not examined in detail, it could be the subject of future research.

Conclusion

This chapter has provided insights into the development of a responsible form of integrated rural tea tourism whereby local and international visitors can gain an appreciation of tea cultures and production in natural settings. In the cases considered, small-scale tea-related tourism projects have served as a form of community and rural development for the areas of tea production while enhancing local livelihoods. Tea-related tourism here is considered as a type of rural tourism. As Japan's rural areas where tea is grown, including the tea villages studied, are suffering from a decreasing population, small-scale tea tourism initiatives can potentially contribute to social regeneration and improvement of community life.

In the case of the tea villages highlighted in this chapter, applying the concepts of responsible and integrated rural tourism, enables both the hosts and the guests to interact in authentic settings, enhancing the visitor experience while contributing to the sustainable livelihoods of the community members. This type of tourism also contributes to and enhances the rural culture and landscapes of Japan. From a wider perspective, it could be argued that the revitalization of rural areas through tea tourism could contribute to the preservation of the unique socio-ecological landscape of rural Japan called *Satoyama* (Takeuchi et al., 2003). Created by centuries of human–nature interactions, tea tourism is predominantly taking place in *Satoyama* rural production landscapes located at the edge of mountains. However, the *Satoyama* cultural landscape is currently suffering from the effects of depopulation such as the loss of human and cultural functions. In this light, tea tourism has the potential of being adopted as a conduit for revitalizing *Satoyama*, which is a testimony to the Asian philosophy of harmony between humans and nature.

While this chapter has considered the development of responsible rural tourism in Japan's tea villages from a supply side, future research could look at visitor profiles and experience in order to gain deeper insights into how rural tea-related tourism is developing in the Japanese tea producing countryside. This chapter provides not only a comparative case study of responsible rural tourism endeavours in three tea villages in Japan, but an extension of the prior literature on tea tourism in rural settings. It proposes that tea tourism in rural settings, when implemented in an evolving and integrated manner, can be a viable form of responsible rural tourism. Lessons from this research could be applied to other tea producing villages in many of the Asian countries that produce tea.

References

Arahi, Y. (1998) *Rural Tourism in Japan: The regeneration of rural communities.* Extension Bulletin – ASPAC, Food & Fertilizer Technology Center. No. 457, 14 pp.

Aylward, E. and Kelliher, F. (2009) Rural tourism development: Proposing an integrated model of rural stakeholder network relationships. IAM conference, 2–4 September.

Beeton, S. (2005) The case study in tourism research: A multi-method case study approach. In B.W. Ritchie, P.M. Burns and C.A. Palmer (eds) *Tourism Research Methods: Integrating Theory with Practice* (pp. 37–48). Wallingford: CABI.

Bixia, C. and Zhenmian, Q. (2013) Green tourism in Japan: Opportunities for a GIAHS pilot site. *Journal of Resources and Ecology* 4 (3), 285–292.

Brotherton, B. (1999) *The Handbook of Contemporary Hospitality Management Research*, Chichester: Wiley.

Cawley, M. and Gillmor, D.A. (2008) Integrated rural tourism: Concepts and practice. *Annals of Tourism Research* 35 (2), 316–337.

Chee-Beng, T. and Yuling, D. (2010) The promotion of tea in South China: Re-inventing tradition in an old industry. *Food and Foodways* 18 (3), 121–144.

Cheng, S., Hu, J., Fox, D. and Zhang, Y. (2012) Tea tourism development in Xinyang, China: Stakeholders' view. *Tourism Management Perspectives* 2, 28–34.

Fernando, P., Kumari, K. and Rajapaksha, R. (2017) Destination marketing to promote tea tourism socio-economic approach on community development. *International Review of Management and Business Research* 6 (1), 68.

Fernando, P., Rajapaksha, R and Kumari, K. (2016) Tea tourism as a marketing tool: A strategy to develop the image of Sri Lanka as an attractive tourism destination. *Kelaniya Journal of Management* 5 (2), 64–79.

Fisher, A. (2010) *The Way of Tea: Reflections on a Life with Tea*. North Clarendon, VT: Tuttle Publishing.

Hashimoto, A. and Telfer, D.J. (2010) Developing sustainable partnerships in rural tourism: The case of Oita, Japan. *Journal of Policy Research in Tourism, Leisure and Events* 2 (2), 165–183.

Higashi Songi Green Tea Tourism Association (n.d.). See http://greentea-homestay.com/#association (accessed 1 June 2018).

Ikkyu Tea (n.d.) Jisaku Ooyama. See https://ikkyu-tea.com/en/green-tea-farmer-sonogi-ooyama (accessed 15 June 2018).

Japan Tea Export Council (n.d.) Current Issues and Situation. See http://www.nihon-cha.or.jp/export/english/problem.html (accessed 15 July 2018).

Jolliffe, L. and Aslam, M. (2009) Tea heritage tourism: Evidence from Sri Lanka. *Journal of Heritage Tourism* 4 (4), 331–344. DOI: 10.1080/17438730903186607.

Jolliffe, L. (2007) *Tea and Tourism: Tourists, Traditions and Transformations*. Clevedon: Channel View Publications.

Knight, J. (1996) Competing hospitalities in Japanese rural tourism. *Annals of Tourism Research* 23 (1), 165–180.

Knight, J. (1994) Rural revitalization in Japan: Spirit of the village and taste of the country. *Asian Survey* 34 (7), 634–646.

Kyoto Obobu Tea Farms (n.d.) See https://obubutea.com/ (accessed 1 November 2018).

Lane, B. (1994) What is rural tourism? *Journal of Sustainable Tourism* 2 (1–2), 7–21.

Nao, S. (2017) Osawa Engawa Cafe in Shizuoka: Feel the true Japanese hospitality! *Matcha Travel Magazine*, July 15, 2017. See https://matcha-jp.com/en/895 (accessed 25 July 2018).

Ohe, Y. 7 Kurihara, S. (2013) Evaluating the complementary relationship between local brand farm products and rural tourism: Evidence from Japan. *Tourism Management* 35, 278–283.

Pitelka, M. (2013) *Japanese Tea Culture: Art, History and Practice*. London: Routledge.

Roberts, L., Hall, D. and Morag, M. (2017) *New Directions in Rural Tourism*. London: Routledge.

Robertson, J. (1988) The culture and politics of Nostalgia: Furusato Japan. *International Journal of Politics, Culture and Society* 1 (4), 494–518.

Service Civil International (n.d.) International Volunteering. See https://workcamps.info/icamps/camp-details/camp-11123.html (accessed 10 July 2018).

Takeuchi, K., Brown, R., Washitani, I., Tsunekawa, A. and Yokohari, M. (2003) *Satoyama: The Traditional Rural Landscape of Japan*. Tokyo: Springer. DOI: 10.1007/978-4-431-67861-8.

Tomocchi. (2017) Higashi Sonogi's amazing rice bowl and beautiful tea fields. Dejima Network for Foreigners with Connections to Nagasaki. See http://dejima-network.pref.nagasaki.jp/en/tomocchi/26212/ (accessed 23 March 2017).

Urry, J. (1988) Cultural change and contemporary holiday-making. *Theory, Culture & Society* 5 (1), 35–55.

Zavadckyte, S. (2013) Japanese Tea Tour in Wazuka. *Japan Travel*, July 13, 2013. See https://en.japantravel.com/kyoto/japanese-tea-tour-in-wazuka/5516 (accessed 26 July 2018).

5 Community Garden Experience During the Off-Peak Tourism Rainy Season at Hoi An Heritage Town, Vietnam

Thu Thi Trinh

Introduction

Tourism is often encouraged by governments or development agencies as a way to support traditional economic and social activities. It is also noted that any type of tourism, community characteristics and nature of host–guest interactions would be included in any list of factors affecting the nature of tourism impacts. Tourism brings with it inevitable positive and negative impacts that arise from the interrelationship between host communities, visitors and the natural environment (McKercher, 1993). Likewise, long admired for their natural beauty and tranquillity, gardens have long been popular tourist destinations. More recently garden activities have been incorporated as an aspect of rural farming tourism, providing an additional source of revenue for gardens/farms that support traditional agricultural practices while increasing both income and the quality of life for farming families. This has become a feature of tourism in many Asian countries such as Taiwan, Sri Lanka, Malaysia and, as in this study, Vietnam (Goh, 2016; Pushpakumara, 2016).

In recent years, the tourism industry of Vietnam has developed rapidly and has become a major economic growth driver with the strong support of government policies. The tourism sector has been identified as a driver of various economic activities and the sector is expected to contribute towards economic growth and distribution of wealth in the nation. As a nation rich in history and agricultural culture, the heritage rural tourism sector is emerging as an important sector of the tourism industry. Vietnam Central has one of the best cultural heritage rural destinations in the region and most of these world-class destinations are set in the rural landscape including world heritage site Hoi An town in Quang Nam province. The Central is the area that is usually rainy and the typhoon season in the Central Coast usually occurs from September to December. The rains last several days, with constantly heavy showers with the typhoon causing floods sometimes. This chapter particularly considers rural community garden tourism in a heritage town setting during the rainy season.

Some studies have shown that garden visitors are often motivated by recreational and leisure interests (Navrátil et al., 2016). Others have revealed that tourists felt that such visits have permitted learning, stress relief and relaxation, and such experiences may also provide insights into the lives of other cultures (Goh, 2016). The practice of visiting community gardens during the rain and flood season has been little researched compared with other types of attractions and open spaces. In particular, while a number of authors argue that environmental education, environmental responsibility and community culture preservation should be a beneficial and critical component of responsible tourism experiences (Nair & Azmi, 2008), there has been a lack of research as to what is being gained by the individual tourist at a specific garden tourism attraction.

This chapter focuses on tourists' experience at community organic gardens during the off-peak tourism period that occurs during the rainy season in Hoi An heritage town and attempts to understand the needs, wants and beneficial perceptions of tourists currently visiting the place. The chapter is based on 33 in-depth interviews with both domestic and international tourists conducted during the rainy season. Community garden activities can play a significant role in bridging formal and informal learning, particularly for urban participants, as it enhances relational experiences about cultural integration, food security, social interactions, community development and environmental concerns. A sense of responsibility was found to emerge among stakeholders who proposed that organic gardens should be managed to provide and promote relaxation while maintaining a healthy, interesting and diverse collection of gardens that enables visitors to experience new botanical learning journeys. Additionally, this chapter opens the discussion on monsoon tourism and responsible tourism as it may play a pivotal role in the future economy for local residents and a sense of responsibility for ecological habitus and climate change.

Gardens and Experiential Tourism

Increased attention has been given in tourism literature to experiential perspectives, especially with regard to garden tourism. Experiential elements of tourism, in terms of feelings, sensations and consumer thoughts, are an important topic for investigation as tourists are increasingly concerned with not just being 'there', but with participating, learning and experiencing the 'there' they 'visit' (Willson & McIntosh, 2007). By revealing how tourists react to, or benefit from, the experiences gained through their subjective associations with places, an experiential perspective can provide important information for product development and marketing, and provide a useful analytical perspective for successful service encounter management (Prentice, 2001). A garden is understood as 'a piece of ground adjoining a house, used for growing flowers, fruits, or vegetables', or representing 'ornamental grounds laid out for public enjoyment and recreation'. It is also an activity. One can become an active participant by working in a garden as a 'gardener' or by having 'gardened'. The tradition of visiting gardens is likewise not a product of modern or post-modern society, but has a longer historical tradition (Lipovská, 2013). Public gardens, parks and botanic gardens attract a substantial number of domestic and international visitors throughout the world (Connell & Meyer, 2004). The tourists in this group are quite diversified and certain gardens can be distinguished and commercialized through tourism (Benfield, 2013).

Previous research has explored the personal and affective dimensions of tourists' experiences in garden tourism. As an example, Navrátil et al. (2016) identified three motivating factors for tourists in making their choice for chateau gardens at Central and Eastern European perspectives and they are mutually complementary such as knowledge and relaxation while other types such as amusement are not used as much. In other words, the main motive for visiting gardens, particularly within the botanical gardens and city parks, is the possibility to stay in a pleasant natural environment and to enjoy outdoor beauty (Benfield, 2013). Another reason is to visit a nice environment, as well as enjoy tranquillity (Connell, 2004). A third reason is to enjoy the natural beauty of the location and a fourth reason is to relax (Lin et al., 2017). The motives of learning and discovering were also studied (Ballantyne et al., 2009), as botanical gardens include education as one of their objectives (Ward et al., 2010). Nevertheless, education alone is not usually the main motivator to make a visit, but it is considered as 'a very important outcome' (Benfield, 2013: 173). Other important factors that influence the turnout in parks are their history, a link with an eminent personage or event and the presence of blockbusters in the garden as well (Benfield, 2013).

To some extent, as 'flagship' attractions, such gardens may also offer a sustainable form of tourism development, with positive impacts on the tourism economy in their local region (Sharpley, 2007). Botanic gardens in particular are traditionally associated with environmental conservation and education, and they typically contain collections of plants for education, scientific purposes and display. In particular, botanic gardens have the potential to provide informal learning experiences that not only promote the importance of plants, habitats and conservation, but also influence the values, attitudes and actions of their visitors. Throughout the world, botanic gardens are starting to take greater responsibility for educating the public about global environmental change and conservation issues (Mintz & Rode, 1999). These gardens are usually informal and aesthetically pleasing, and are thus particularly well-placed to showcase the inter-relationships among plants, animals and humans and to explain how the different components are inextricably linked.

There are the physical characteristics of space and that space and the interpretive process and presentation may draw the visitor into an experience that exceeds any initial expectation in which the site/place sets out to involve the visitor through patterns of inviting reactions and through interpretation and visitor involvement (Trinh & Ryan, 2017). The most important aspects of the gardens/ parks are the trees, plants, the greenery, the flowers, and fresh air; they are part of the essentials that are offered in the quality of a park/garden (Connell & Meyer, 2004). However, it is argued that a garden or park is visited simply because it is situated in a visited area. This fact was not only found in the case of the botanical gardens (Ballantyne *et al.*, 2009), but it was also true in chateaux in general (Navratil *et al.*, 2016). Additionally, visits to these gardens are mostly a seasonal activity, when the maximum number of visits is done from April to September and the development of indoor expositions has created year-round opportunities. For example, the exhibitions of numerous tropical blooming orchids during winters in the temperate zone of the Northern Hemisphere, as a focus of the demand for garden tourism (Benfield, 2013). Despite this, there remains a notable gap in the literature about the more individualized, personal meanings and sense that tourists place on gardens in a rural heritage setting during a rainy time/season. There has been little attention given to the experiential nature of gardens. In seeking to contribute to an increased understanding of the experiential nature of tourism encounters, the current chapter presents an exploration of the experiential nature of garden products in a heritage town.

Community Organic Vegetable Gardens in Hoi An World Heritage Ancient Town

A case study of private organic vegetable garden visits in Hoi An heritage town was an appropriate methodological approach, as case studies allow for research to be based in a practical real-life situation and the in-depth results obtained and the discussions are pertinent to real-world situations (Robson, 2002). Set in a quiet environment, Hoi An is surrounded by peaceful villages focusing on vegetables, carpentry, bronze making, ceramic etc. In particular, research was conducted in the organic gardens of the villages of Cam Ha and Cam Thanh Communes, Hoi An, Quang Nam province; each commune can be justified for inclusion on a number of grounds as a subject of research. First, as a United Nations Educational, Scientific and Cultural Organization (UNESCO) inscribed World Heritage site, the town possesses importance as a tourism, cultural and heritage centre. Further, reports from the Youth News of Vietnam highlight that vegetable garden farms in Hoi An are famous for their earthy aroma as the farmers use organic farming methods complemented by a combination of natural factors such as sunlight, water and soil in the area. Additionally, there are statements from the Vice Chairman of Hoi An heritage site that emphasize the positive contribution of organic farming activities to the economy of the village as it attracts hundreds of tourists on a daily basis: 'The city aims to maintain areas reserved for paddy fields to promote tourism-focused service agriculture' …. 'It provides local farmers with income not only from the sale of vegetables but also from the tourist industry'. Its claim to authenticity at these gardens is also based on its heritage rural surroundings and in many such cases, it has become a fusion between the need to entertain tourists on farming and a portrayal of a heritage wherein local people claim back a history that demarcates them from dominant agricultural cultures.

The rainy season in Hoi An town begins in August and lasts until December. Many people consider travel in the rainy season as taboo, but perhaps for Hoi An this is an exception. This research aimed to explore the motivations and perceptions of visitors visiting these gardens. The issue also lies on whether organic garden tourism during the monsoon provides a quality travel experience that promotes conservation of the natural environment and offers opportunities and benefits for local rural communities. The mode of study adopted in this research is discussed in the next part of this chapter.

Methods

Generally, visiting a tourist attraction is likely to involve a flow of experiences which seem to describe an evolutionary experience of place that is likely to move some visitors from a position of relatively shallow interest to an almost emotional position due in some part to the role of the aesthetics of the landscape (Trinh & Ryan, 2016). In exploring the experiential aspects provided by vegetable gardens in the selected Hoi An heritage town, the research sought to address two main questions: Why do people visit gardens and comparable heritage attractions during the rainy season? What do visitors actually expect and perceive from their visit? The initial purpose of the research was to assess the extent to which a visit generates feelings about a place that may differ from those of their original feeling from a visit to a garden located in a heritage site, and to use an inductive approach by which concepts would be sought within the data set, rather than seek to test any initial theoretical concept. Specifically, particular attention has been paid to the ways in which visitors' experiences of garden destination areas translate into a more detailed understanding and appreciation of local culture or history (Taylor, 2001). Also of interest will be the change in visitors' perceptions and attitudes towards hosts, thereby potentially minimizing the negative impacts of tourism on host communities.

As an exploratory piece of research, the exercise sought to generate data using a micro-ethnographic approach. An initial exploratory stage was conducted before the more focused round of data collection that forms the major component of the study. The site was visited to observe tourists' behaviours, and a diary was kept during each day of observation and subsequent data collection during the rainy season. Using qualitative methods, this study explores the nature of demand for garden tourism in a heritage setting of exotic landscapes and architecture with particular attention to the appreciation gained by visitors during the rainy season (from late August to December). The rain and typhoon season is a characteristic of tropical countries, including Vietnam, one of the tropical monsoon countries in the world. The typhoon season in Hanoi and other Northern provinces is not the same as in the South; it does not last many months, just alternates between periods of sunny and hot weather (in the summer). The typhoon season in North Central Vietnam occurs from July to October and in the South Central Coast (of this research), usually from September to December. The rains last several days, causing widespread and extensive flooding.

Convenience sampling was adopted in terms of the fact that interviews were conducted at comfortable cafés or restaurants nearby/next to gardens where visitors take a rest after their visit and was also purposeful with domestic and international tourists being deliberately selected. The interview focused on an exploration of the more personal context of visiting, rather than a quantitative measurement of the experiences gained. Many researchers believe that while the designation of Hoi An town as World Heritage Site can be a catalyst to rapid tourism development, the introduction of large volumes of tourists from different socioeconomic and cultural backgrounds potentially poses a major threat to the sustainability of the sites and its rural neighbouring communities. The main questions asked included: Where are you from? What are your main motivations for this visit? Why do you visit the vegetable garden on rainy days/typhoon season? What experiences have you gained from these visits? Hence for these reasons, a phenomenographic approach (Marton, 1981) was adopted where key questions comprised prompts such as 'What do you mean by that?' or 'Can you tell me a little more why you think so/say so?' Consequently, in any series of interviews, commonalities of themes tend to emerge relatively quickly, particularly under careful probing and the use of common patterns of questioning.

Discussions of Findings

This chapter explores community vegetable garden visits in the monsoon season. Three themes emerged from this Hoi An rural-based research.

Repeat visitors, nostalgia, a sense of place

The demographic profile, nationality, gender, travel style and age of respondents were recorded. Interestingly, more than 80% of the sample was repeat visitors. When asked about the main motive for revisiting these organic vegetable gardens, the majority of repeat visitors said that it was simply because of a close attachment to this site, a sense of place on typical rainy days much influenced by the beauty of the rainy season, and a perception of the rain and the rural atmosphere as really pleasantly unique, 'hard to express feeling but I love this rural place as I find something new each time'. Tourism experiences are indeed of a basically visual nature and these views suggest that tourists may have shallow motivation prior to the visit but then re-evaluate their experience as a result of what they have seen and revisit the garden. This process of re-evaluation can be attributed to emotional attachment or sensory appreciation.

> 'Pick up the umbrella in the rain, take a walk or sit in a certain café with a cup of hot tea, listen to the melody of a light music ... watch the rain covered garden ... green plants on a rainy day is sure to be one of the best experiences that visitors can hardly find somewhere else... various feelings of different visits for one place ... When it rains - usually mid-afternoon – you know you've been rained on, but then the sun comes out, the gardens become picturesque'.

More than a half of the repeat respondents (73%) claimed that they wanted to personally see/view/observe community green garden, and experience organic farming practice in the rainy season. Some have interest in buying organic vegetables (35%) and their focus seems to involve/ attachment to local tourism-related activities (80%) for a pleasant rainy day out. Eating local food/dishes cooked with local organic herbs collected from on-site organic gardens, connecting with visitor's interest in gardening experiences they have had before or had in their own farming style, is also likely to be popular (60%). In particular, the majority of international visitors (72%), mainly the older ones, shared how they used to spend their free time gardening as a free-time activity, far less than watching television: 'We visit gardens as part of visiting historic sites or to learn about gardens and gardening'. Hence, the beneficial experiences gained at the organic gardens were found to be affective in nature as well as cognitive. Likewise, a survey in the Adelaide Botanic Gardens by Crilley and Price (2005) found that although 57% of respondents cited 'viewing plants' as one of the three main reasons for visiting, only 15%

were motivated by the desire to 'learn about plants'. Indeed, it is generally accepted that the majority of visitors to botanic gardens do not come to learn *per se* (Darwin- Edwards, 2000).

On the other hand, the initial experience of novelty may become a subsequent visit based on nostalgia. Ryan (2002) argues that if the core of tourism is the experience of visit to a place, then repeated visits to a place are not repeat experiences, because each visit is built upon prior learning. Similarly, McKercher (2002) argues that the 'specialized cultural tourist' focuses his or her efforts on one or a small number of geographical sites or cultural entities. As such, this type of tourist revisits a particular garden or rural town in search of a deeper understanding of that place or in search of exemplars of a specific kind of art or simply a recall of childhood or the farming/gardening habits on farms in a rainy season. A further consideration is that a tourist may revisit an activity rather than a place as some tourists state that they revisit organic gardens to simply experience boat or bicycle riding and hence comfortably visit rural gardens nearby.

Transition to tourism during the off-peak tourism rainy season: responsible tourism for a new approach to monsoon garden tourism

Climate and weather are not only important factors in the decision making of tourists, but they also influence the successful operation of tourism businesses. Climate is either the main tourism resource, for example, in the case of beach destinations (Kozak *et al.*, 2008), or acts as a facilitator that makes tourism activities possible and enjoyable (Gomez, 2005). While tourists might expect certain climatic conditions when they travel to a place, they will experience the actual weather, which might deviate quite substantially from the average conditions (Becken, 2010). The tropical monsoon climate can bring much hardship and the monsoon season is commonly considered 'off-peak' tourism as the tropical monsoon climate is characterized by heavy seasonal rainfall, high temperatures and high humidity, all of which badly affect normal living.

Yet, the increase in tourists at rural Hoi An gardens during the rainy season is a 'phenomenon'. Hoi An town now is offering different types of cruises, motor boat rides or even a boat trip to enjoy and explore the scenic beauty of the monsoon climate. This is an attractive way of holding the attention of both domestic and international tourists.

During the monsoon season, many flowers and vegetables bloom and enhance the beauty of these rural gardens. Visitors' feelings and comments gathered from interviews described their garden visits as 'Unique heritage setting in the rain', 'the romance of the monsoon', 'green space', 'abundant rains', 'natural/semi-natural areas covered by vegetation'. These were perceived as a 'unique' tourist attraction. Tourists may visit many villages, see the real-life style of the locals in the rainy season, and enjoy the different lively rural cultural programmes/activities. The data suggest that the visual, aural, smells, sound and touch of vegetable gardens create sensory links to current appreciation of place and subsequent memory of attendance. These in turn may link to the cognitive which is rural environment awareness and climate change towards organic gardens within the monsoon tourism season.

Through these preliminary insights, the in-depth interviews revealed only three visitors who were likely to be dissatisfied with the activities undertaken. These international respondents could be classified as 'purely gardeners and conservationists' who have concerns about farming/garden settings potentially affected by tourism activities. A respondent from the UK even questioned the extent to which the host has diversified the organic gardens to suit tourism. In particular, the current setting and facilities may not provide sufficient quality for visitors during the monsoon season and for some big-size tours. Striking themes/phrases such as the following were expressed: 'wet/ slippery roads', 'rough roads', 'uncontrolled tour sizes', 'bumpy rides', 'not being able to get to garden/gardening practices', 'concern about impact of diversification on organic farming into tourism', 'too commercial', 'need more guides to talk on organic garden farming', 'need better quality farming facilities and garden tourism services'. A sense of responsibility was found to emerge among visitors who proposed that organic gardens be managed to provide and promote relaxation while maintaining a healthy, interesting, and diverse collection of gardens that enables visitors to experience new botanical learning journeys. As such, this chapter provides for discussion on monsoon tourism and responsible tourism, given that it may play a pivotal role in the future economy of local residents along with an enhanced sense of responsibility for climate change.

In-depth interviews were carried out to explore the long-term benefits and expectations of tourists' garden visits/ viewing tourism attractions and, in particular, to identify whether positive awareness and action towards the environment results from such experiences. Findings revealed tourists' concern for degradation of rural infrastructure due to the flood at the peak of monsoon tourism. In particular, rural narrow roads to gardens in the rainy season could be potentially damaged as the tourist numbers have increased, and this may affect resident's lives. Thus, there is a need for on-site interpretative media in fostering visitor appreciation of resources, environmental awareness and behaviour. In short, more visitor studies are needed to understand how the human dimension of sustainable rural monsoon tourism management can ensure the preservation of the natural farming environment and organic gardens.

On the other hand, visits at rural community organic gardens may have the potential to provide informal learning experiences that not only promote the importance of plants, habitats and conservation, but also influence the values, attitudes and actions of their visitors. An increased understanding of the importance of rural gardens and farming in the tourist experience could be valuable in influencing future policy regarding the preservation of the farming environment within a heritage setting. To do this effectively, there is a dire need for a well-designed interpretation that communicates this importance.

Conclusion

It is noted that despite the growing body of garden tourism research and the vulnerable standing of tangible heritage in rural areas, there is limited evidence in the literature on the linkage between garden tourism and the preservation of tangible rural heritage and sustainability. In the tropical country of Vietnam which experiences regular typhoons, floods and a long rainy season, it is likely that further research on garden tourism and monsoon tourism should be carried out in the light of responsible tourism. Today, small-scale rural tourism, in particular, garden and farming tourism is in the process of becoming an important activity that is expected to promote employment, vitality and sustainability of rural communities, given that private gardens can become powerful cooperative tools of community creation. As such, tourism is part of the shift in the economic base of rural societies. Tourists want to see the tangible heritage of a site, but may also want to experience 'the backstage' of intangible, lived agricultural/farming culture associated with that site. As rural farming and gardening have come under increasing pressure to diversify into tourism, pluriactivity has become a fact of life, creating important pillars supporting farming and making it possible for farms, that otherwise would have been forced to disappear, to stay in business (Kinsella *et al.*, 2000; Ploeg *et al.*, 2000).

Additionally, the romance of the monsoon, with its abundant rains combined with the lush green gardens of Hoi An heritage town, can be promoted as a 'unique tourist attraction', introducing a new attraction for the tourists who come to explore the rural beauty of the rainy season and in that process generate employment opportunities for the local people. Garden and monsoon tourism has huge potential to create a remarkable change in the economic growth of Vietnam. One of the key questions relates to adaptation on the part of residents who are exposed not just to more tourism, but to potential uses of natural and historical resources as well as changing land use patterns. Also the commercialization of past traditions as more festivals and events are created for tourists to experience, that allegedly are based in a past life style, and synchronize it with political perquisites that favour economic development.

Vietnam is located in a region which is 25 times more susceptible to natural disasters than Europe (Asian Development Bank, 2013). The monsoon is a common phenomenon across Asia and many countries in the region are now leveraging on the monsoon to develop customized tourism experiences, with varying degrees of success. Torrential and prolonged rain with high humidity during the monsoon coupled with human activities such as the organic farms and gardens in Hoi An have created a uniquely Vietnamese sense of place and a romantic gaze (Urry & Larsen, 2011). The place-making process shaped by the monsoon is common not only in Vietnam but in many parts of Asia that have a tropical climate – and the resulting sense of place is a manifestation of Asia's acceptance of its susceptibility to natural disasters and the need to maintain a harmonious relationship with its fragile physical environment. Rather than regarding natural calamities and phenomenon such as the monsoon as an impediment, the Hoi An case study shows how it could be celebrated and transformed into a responsible rural tourism experience that is based on a uniquely Asian form of rurality.

While monsoon tourism in Bangladesh has been researched and introduced as a new phenomenon and is a new aspiration for Bangladesh tourism (Chowdhury & Kassem, 2014), major gaps remain in the collective knowledge about the potential of Vietnam's monsoon tourism within garden visits of a heritage setting. In particular, further research is needed on whether such diversification of the monsoon season and organic gardens into sustainable tourism serves both farming practice and visitors' quality experiences and whether such development enhances cultural and environmental sustainability or responsible tourism. Further research is certainly warranted on responsible tourism at rural community garden attractions to make a significant impact on both tourist experience and lives of residents during the so-called off-peak rainy tourism season.

References

Asian Development Bank (2013) *The Rise of Natural Disasters in Asia and the Pacific*. Manila, Philippines: Asia Development Bank Publications.
Ballantyne, R., Packer, J. and Hughes, K. (2009) Tourists' support for conservation messages and sustainable management practices in wildlife tourism experiences. *Tourism Management* 30 (5), 658–664.
Becken, S. (2010) The importance of climate and weather for tourism: Literature review. See https://pdfs.semanticscholar.org/dd26/316707021167c56e22c0d272014346078cc7.pdf (accessed 5 February 2018).
Benfield, R. (2013) *Garden Tourism*. Wallingford: CABI
Chowdhury, S.A. and Kasem, N. (2014) Monsoon tourism: A new aspiration for Bangladesh tourism industry. *Journal of Tourism, Hospitality and Sports* 2 (1), 2312–5187.
Connell, J. and Meyer, D. (2004) Modelling the visitor experience in the gardens of Great Britain. *Current Issues in Tourism* 7 (3), 183–216.
Crilley, G. and Price, B. (2005) *The Adelaide Botanic Gardens Visitor Service Quality Survey*. Centre for Environmental and Recreational Management, University of South Australia, Adelaide.
Darwin-Edwards, I. (2000) Education by stealth: The subtle art of educating people who didn't come to learn. *Roots* 20, 37–40.

Goh, H.C. (2016) The user's perceptions of Perdana Botanical Garden in Kuala Lumpur. *Journal of Design and Built Environment* 16 (1), 27–36.

Gomez, M.M. (2005). Weather, climate and tourism: A geographical perspective. *Annals of Tourism Research* 32 (3), 571–591.

Kinsella, J., Wilson, S., De Jong, F. and Renting, H. (2000) Pluriactivity as a livelihood strategy in Irish farm households and its role in rural development. *Sociologia Ruralis* 40 (4), 481–496.

Kozak, N., Uysal, M. and Birkan, I. (2008) An analysis of cities based on tourism supply and climatic conditions in Turkey. *Tourism Geographies* 10 (1), 81–97. 10.1080/14616680701825230

Lin, B.B., Gaston, K.J., Fuller, R.A., Wu, D., Bush, R. and Shanahan, D.F. (2017) How green is your garden? Urban form and socio-demographic factors influence yard vegetation, visitation, and ecosystem service benefits. *Landscape and Urban Planning* 157, 239–246.

Lipovská, B. (2013) The fruit of garden tourism may fall over the wall: Small private gardens and tourism. *Tourism Management Perspectives* 6, 114–121.

Marton, F. (1981) Phenomenography: Describing conceptions of the world around us. *Instructional Science* 10 (2), 177–200.

Mckercher, B. (2002) Towards a classification of cultural tourists. *International Journal of Tourism Research* 4 (1), 29–38. 10.1002/jtr.346.

McKercher, B. (1993) The unrecognized threat to tourism: Can tourism survive 'sustainability'? *Tourism Management* 14 (2), 131–136.

Mintz, S. and Rode, S. (1999) More than a walk in the park? Demonstration carts personalize interpretation. Roots: Botanic Gardens conservation. *International Education Review* 18, 24–26.

Navrátil, J., Kučera, T., Pícha, K., White, V.L., Gilliam, B. and Havlíková, G. (2016) The preferences of tourists in their expectations of chateau gardens: A Central and Eastern European perspective. *Journal of Tourism and Cultural Change* 14 (4), 307–322,

Nair, V. and Azmi, R. (2008) Perception of tourists on the responsible tourism concept in Langkawi, Malaysia: Are we up to it. *TEAM Journal of Hospitality and Tourism* 5 (1), 27–44.

Ploeg, J., Renting, H., Brunori, G., Knickel, K., Mannion, J., Marsden, T., Roest, K., Sevilla-Guzman, E. and Ventura, F. (2000) Rural development: From practices and policies towards theory. *Sociologia Ruralis* 40 (4), 391–408.

Prentice, R. (2001) Experiential cultural tourism: Museums and the marketing of the new romanticism of evoked authenticity. *Museum Management and Curatorship* 19 (1), 5–26.

Pushpakumara, K.G.J. (2016) Valuing ecosystem services, biodiversity and their contribution to farmer's livelihood with potential for future benefit sharing mechanism of forest garden farming systems of three different agro-ecological zones in Sri Lanka. *37th Asian Conference on Remote Sensing, ACRS 2016*, Vol. 1, pp. 652–664.

Robson, C. (2002) *The Analysis of Qualitative Data*. London: Blackwell.

Sharpley, R. (2007) Flagship attractions and sustainable rural tourism development: The case of the Alnwick Garden, England. *Journal of Sustainable Tourism* 15 (2), 125–143.

Ryan, C. (2002) *The Tourist Experience*. London: Continuum.

Taylor, E. (2001) Positive psychology and humanistic psychology: A reply to Seligman. *Journal of Humanistic Psychology* 41 (1), 13–29.

Trinh, T.T. and Ryan, C. (2017) Visitors to heritage sites: Motives and involvement: A model and textual analysis. *Journal of Travel Research* 56 (1), 67–80.

Urry, J. and Larsen, J. (2011) *The Tourist Gaze 3.0* (3rd edn). London: Sage Publications.

Ward, C.D., Parker, C.M. and Shackleton, C.M. (2010) The use and appreciation of botanical gardens as urban green spaces in South Africa. *Urban Forestry and Urban Greening* 9 (1), 49–55.

Willson, G.B. and McIntosh, A.J. (2007) Heritage buildings and tourism: An experiential view. *Journal of Heritage Tourism* 2 (2), 75–93.

6 Responsible Rural Tourism Initiatives and Local Community Development in Kerala, India

Anu Treesa George, Terry DeLacy and Min Jiang

Introduction

Tourism is an important industry which generates foreign exchange earnings and local income, as well as contributes to social harmony (United Nations World Tourism Organization [UNWTO], 2007). It is a promising and prominent industry which contributes to economic growth and development in developing countries. As per the UNWTO General Assembly, 19th Session 'Tourism Towards 2030' and the Annual Report on Tourism Highlights 2014 edition, the strongest growth in international tourist arrivals will be seen in Asia and the Pacific, where arrivals are forecasted to reach 535 million in 2030 (Sharpley & Telfer, 2014; UNWTO, 2011). Tourism is a multidimensional industry and a new mantra for socio-economic development in the Asia-Pacific region, especially for developing nations.

Responsible tourism reflects sustainable tourism, a concept which is framed on the responsibility aspects of stakeholders in tourism (Debicka & Oniszczuk-Jastrzabek, 2014; Leslie, 2012). Responsible tourism is about making better places for people to live in and better places for people to visit (Cape Town Declaration, 2002). It focuses on inspiring visitors, local community and stakeholders in the tourism industry to take greater responsibility for making tourism sustainable. The UNWTO in the 'Tourism in the Green Economy' report recognized sustainable approaches as that which 'aspire to be more energy efficient and more climate sound, consume less water, minimize waste; conserve biodiversity, cultural heritage and traditional values; support intercultural understanding and tolerance; generate local income and integrate local communities with a view to improving livelihoods and reducing poverty' (UNEP, 2010: 2).

This chapter provides a brief account of the background of responsible tourism and the positive impacts of innovative responsible tourism initiatives in Kerala, India. Current practices of responsible tourism initiatives and how it influences the local community are reviewed. The chapter then reports on detailed primary research of responsible rural tourism activities in Kerala, focusing on aspects of economic, social and environmental impacts along with local community development.

Background

Responsible tourism is a local movement, being developed by different people in different ways around the world based on different local issues. It embraces all types of tourism and focuses on results (Debicka & Oniszczuk-Jastrzabek, 2014; Leslie, 2012). The original definition of sustainable development was provided by the Brundtland Commission in *Our Common Future* as 'Sustainable development is a development that meets the needs of the present without compromising the ability of future generations to meet their own needs' (Brundtland, 1987: 29; Clarke, 1997). Sustainable tourism is tourism and associated infrastructure that, '… both now and in the future operate within natural capacities for the regeneration and future productivity of natural resources; recognize the contribution that people and communities, customs and lifestyles, make the tourism experience; accept that these people must have an equitable share in the economic benefits of local people and communities in the host areas' (Eber, 1992: 3). Sustainable tourism is 'Tourism that takes full account of its current and future economic, social and environmental impacts, addressing the needs of visitors, the industry, the environment and host communities' (United Nations World Tourism Organization, 2013).

In the late 1980s, the concept of responsible tourism contributed to a new dimension by encouraging visitors to take responsibility for achieving sustainable development and identifying local socioeconomic issues and tackling them. Since then efforts have continued to explore the concept by looking through the prism of responsibility and identifying the various factors that can contribute to economic and cultural benefits of tourism (Spenceley, 2012). Researchers have studied responsible tourism, investigating how it can enhance social sensitivities and ethical practices in tourism destinations (Debicka & Oniszczuk-Jastrzabek, 2014; Frey & George, 2010; Goodwin & Francis, 2003). From a different point of view, responsible tourism is not a niche tourism product or brand but explains the ways responsible visitation can accelerate and foster the sense of responsibility in visitors and local people (Ebitu, 2010; Harrison & Husbands, 1996).

The Cape Town Conference was organized by the Responsible Tourism Partnership and Western Cape Tourism as a side event preceding the World Summit on Sustainable Development in Johannesburg in 2002. The 2002 Cape Town Declaration focused on minimizing the negative economic, environmental and social impacts; enhancing the well-being of host communities and empowering the local people in decision making that affects their lives (Cape Town Declaration, 2002). It is about identifying the locally significant issues and acting to deal with them. In Cape Town, the city council identified seven local priorities: reductions in water and energy consumption and waste; increased local procurement and enterprise development; social development and skills development. The city council and tourism businesses in Cape Town are taking responsibility to achieve these sustainability objectives and progress is being measured against clear agreed criteria (Leslie, 2012). The 'responsible visitor' aims to enjoy the culture, the customs, the gastronomical offer and the tradition of the local people in a respectful way and always tries to contribute to the development of responsible and sustainable tourism (Debicka & Oniszczuk-Jastrzabek, 2014). The word 'responsible' means a way of being truthful about the fact that all tourism has impacts, which is often difficult to measure. Responsible travel might be a more apt term to describe the current 'best case scenario' for tourism. It means a positive shift in visitors' desire to conserve and be more conscious about resource utilization. Thus, the term responsible tourism focuses on fostering the sense of responsibility by visitors who enjoy the destination and enhance the responsibility of local people in the destination (Frey & George, 2010; Goodwin & Francis, 2003).

Responsible Tourism in Kerala

Kerala is an evergreen state situated on the South-west coast of India between the Arabian Sea in the West and the Western Ghats in the East. It is becoming a major holiday destination with its spectacular and rich 2000-year-old cultural history and scenic beauty. Kerala enjoys unique geographic features and is known as 'God's Own Country' (Thimm, 2016). The region has emerged as the most acclaimed tourist destination in India in the past two decades (Edward & George, 2008). World Travel and Tourism Council (WTTC) selected Kerala as a partner state for enhancing the tourism industry. Figure 6.1 highlights the location of the state of Kerala in the Indian political map and major tourist destinations in the region.

Kerala's unique selling proposition for marketing 'One of the Ten Paradises of the World' is ABC, an acronym for Ayurveda, Backwaters, Culture, and Cuisine (Chettiparamb & Kokkranikal, 2012). It is a beautiful destination with pristine beaches, coastal plains, hill stations, backwaters and the highlands for a relaxing holiday. International tourist arrivals in the year 2017 were 1,091,870, which shows an increase of 5.15% compared to 1,035,630 tourist arrivals in 2016 (Government of Kerala, 2017).

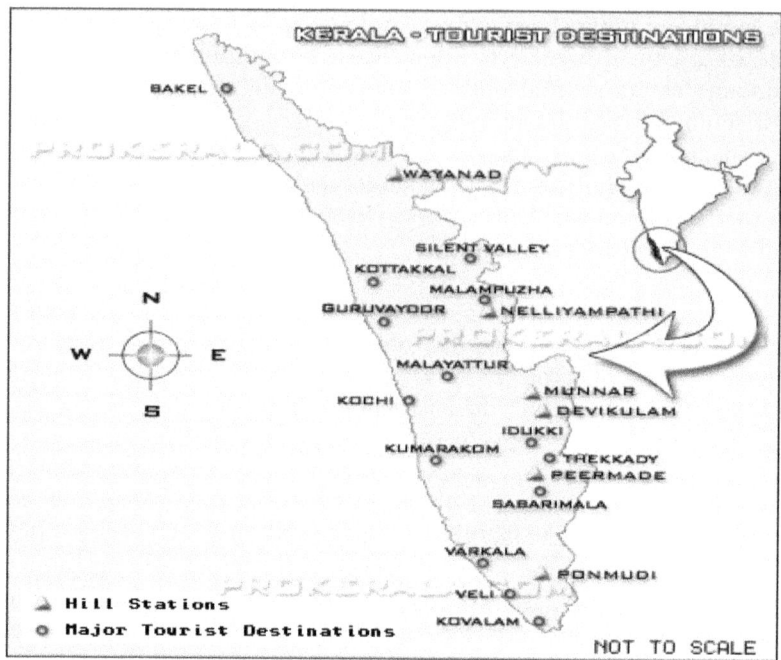

Figure 6.1 Major tourist destinations in Kerala, India
Source: https://www.prokerala.com/kerala/maps/kerala-tourism-map.htm

The unique geographical and cultural features of Kerala provide a wide variety of quality tourism products that not only enhance and enrich tourist experiences but also facilitate the development of Responsible Tourism Initiatives in Kerala (Chettiparamb & Kokkranikal, 2012; Thimm, 2016). Responsible Tourism Initiatives in Kumarakom, one of the tourism areas in Kerala, won prestigious awards of the UNWTO Ulysses Award for Innovation in Public Policy and Governance, the National Award for Best Responsible Tourism Project, the PATA Grand Award for Environment and the World Travel Mart Award for Best Poverty Reduction Project 2017 for the socio-economic impacts on the local community. UNWTO Regional Director for Asia and the Pacific, Xu Jing, congratulated Kerala for its successful 'Responsible Tourism Initiatives' model which could be emulated by other tourist destinations. The projects can be included as part of UNWTO's global programmes and the report can be a case study for other destinations. We are eager to strengthen our relationship and discuss matters of common interest' (India Today, 2017: 3).

In a pilot phase, Responsible Tourism (RT) initiatives were implemented in four tourist destinations namely Kovalam, Kumarakom, Thekkady and Wayanad in Kerala in 2007 under a tag name 'Better Together'. In 2012, the initiative was extended to three other destinations of Kumbalangi in Ernakulam, Vythiri, and Ambalavayal in Wayanad and

Figure 6.2 Responsible tourism organization structure in Kerala, India
Source: Mathew and Kumar (2014).

Bekal in Kasaragod. The major aims of Kerala's Department of Tourism are to promote tourism to generate employment opportunities, reduce poverty and improve the livelihoods of local people. The government had taken initial steps to constitute a State Level Responsible Tourism Committee (SLRTC) to implement the concept of responsible tourism in the destinations. The Kerala Institute of Tourism and Travel Studies (KITTS) was the nodal agency for RT initiatives to provide local self-government and tourism stakeholders with proper training and capacity building (Vijayakumar, 2013). Figure 6.2 shows the RT organizational structure in Kerala.

Case Study: Responsible Tourism Initiatives in Thekkady, Kumily

Study area and research methods

The case study was conducted in Kumily to understand the impacts and expectations of people on responsible tourism initiatives. Representatives of government, tourism industry and local community were interviewed to understand the RT initiatives in the case study destinations. Kumily is the largest Gram Panchayat in Kerala. It is a town in Cardamom Hills near Thekkady, Periyar Tiger Reserve. Kumily is known as the Gateway to Thekkady and is also known as the spice capital of Kerala. To understand the impacts, the study involved 27 tourism stakeholders who included representatives of hotels and restaurants, government officials, responsible tourist destination cells, and local people in Kumily, Kerala. Two focus group discussions were conducted with the responsible tourism spice shop and Community Development Society members of Kumily Panchayat. The research questions focused on understanding the views of local people on current RT policies 'Better Together', positive impacts of these initiatives, identification of the gaps and expectations related to responsible tourism. Table 6.1 provides details about the number of participants included from each category.

Table 6.1 Total number of participants from each category

Representation	Number of participants
Government officials	7
RT coordinators and village life experience packages & local people	13
Safety and security officers	2
Food and accommodation sector	3
Academic field	2
Total	27

Positive impacts of responsible rural tourism initiatives

RT focuses on an approach that ensures benefits to all sectors, local people in tourism destinations and stakeholders (Debicka & Oniszczuk-Jastrzabek, 2014; Okazaki, 2008). Researchers assert the need for RT as it not only benefits the locals by enhancing the well-being of host communities but also makes tourism a more eloquent experience for tourists through meaningful linkages in understanding the cultural diversity of the destination (Blackstock, 2005; Goodwin & Francis, 2003; Reed, 1997). Moreover, RT initiatives must make an effort to include people of all credentials in the tourism sector, including tourists, local business persons or community members at the tourism destinations.

Economic impacts of responsible tourism in the local community

A Destination Level Responsible Tourism Committee (DLRTC) was developed to enhance the relationship between stakeholders in the tourism industry and the local community. After considering the demand for local products and services, the SLRTC decided to implement a State Poverty Eradication Mission that is popularly known as Kudumbashree (Mathew & Kumar, 2014; Venu & Goodwin, 2008). Kudumbashree was appointed as the consultant for RT in the field by the Department of Tourism. Local self-government and Kudumbashree took effective steps for the successful implementation of RT Initiatives. In the initial stages, the Panchayat concentrated on the supply of local products to hotels and resorts in order generate economic benefits for the local community. The consultant conducted a survey to analyse the demand for vegetables in hotels and resorts and drew up an agreement on the regular supply of local products from the local community in the pilot study destinations (Mathew & Kumar, 2014; Venu & Goodwin, 2008).

Linking with the hotel industry

The consultant found that the hotel industry was not procuring local products from the local community. With the intervention of the RT cell, the consultant identified local agricultural products that could be supplied regularly to the hotel industry. In the initial stages, 18 resorts agreed to purchase vegetables, fruits, milk, eggs, honey and tender coconut from local farmers. To ensure quality and a sound pricing mechanism, separate committees were formed to regulate the systems. The communities adopted organic farming with the establishment of an assured market. They expanded progressively with a supply of local items from the local community to the tourism industry. This linkage helped to expand the relationship between community and industry in a win–win partnership, avoided conflicts and improved the quality of life of the people in all aspects (Mathew & Sreejesh, 2017). DLRTC constituted a price fixing committee and a quality committee which became important segments in implementing RT in order to fix the price and assure quality of local products (Mathew & Kumar, 2014). The members of the committees were from the grama panchayat, Kudumbashree, vegetable supplying unit, District Tourism Promotion Council, and purchase staff of hotels. Other members of the committee were the sales tax officer, agricultural officer, chefs of participating hotels and health inspectors. The RT cell also provided an opportunity for the tribal community to open a snack parlour in the Club Mahindra resort in Thekkady. A group of five women prepare and provide traditional Kerala snacks to guests from 3.00 pm to 6.00 pm (Kerala Responsible Tourism, 2008).

The focus group discussion with the Community Development Society revealed that members were satisfied with the support of the government in the production of vegetables, dairy products, eggs etc. Around 170 Kudumbashree units with five women in each group started to cultivate vegetable crops. The RT initiatives were focused on producing vegetables and supplying to hotel groups to generate income to the local community. Hotels and resorts supported the Samruthi shop through purchasing local products. Sales to hotels through Samruthi during January–June 2009 are given in Table 6.2.

In the initial stages of RT initiatives, the supply of products to resorts and hotels was in accord with regulations but key products such as eggs, vegetables, honey, milk etc. were drastically reduced. In the later stages, hotels and resorts that participated in the beginning discontinued due to various reasons including lack of supply from Samruthi group. Interviews with hoteliers and resorts revealed that local suppliers procured vegetables from Kambam market in Tamil Nadu at cheap prices and supplied it to hotel groups with a high margin. Procuring vegetables from the open market was against the concept of RT. After identifying these malpractices, the Department of Tourism, state and destination coordinator supported hoteliers and resorts with strategies to purchase products that were produced locally. Directly or indirectly RT initiatives helped to develop strong links between the tourism industry and local people (Mathew & Sreejesh, 2017).

Table 6.2 Sales through Samruthi during the first half of 2009

Name of Hotel/Resort	Jan-09	Feb-09	Mar-09	Apr-09	May-09	Jun-09
Spice Village	7305	8503	5554		4033	
Shalimar Spice Garden	6626	10,543	11,390	6743	5287	2110
Aranya Nivas	6239	7864	42,882	26,796	61,193	33,330
Kondody Greenwoods Hotel	3600	5842	6795	6423	21,484	9919
Periyar House	3267	6839	9369	3805		
Club Mahindra Resort	2822	8667	21,823	16,621	56,581	18,827
Tree Top Resort	1136	2608				
Muthoot Cardamom County	254	1011	1337			
Lake Palace				3999	7940	8251
Elephant Court Resort			1352			
Total	31,249	51,877	100,502	64,387	156,518	72,437

Note: The figures for January, March, April and June vary within the same source of information.
Source: Report on Responsible Tourism – Phase I, Thekkady, GITPAC-RT Technical Unit, June 2010 (https://www.rtkerala.com/thekkady.php).

Responsible tourism spice shop

Kumily is known for spices such as black pepper, cardamom, cinnamon, clove and nutmeg. The RT cell also identified the scope of a spice shop in Thekkady. Local farmers were asked to supply products such as spices, pickles, turmeric powder, curry powders etc. to the RT spice shop. This move also created employment opportunities for Kudumbashree members and local people. Compared to other spice shops, the profit margin is less in this spice shop. In the focus group discussion, employees in the spice shop highlighted they were homemakers and with the introduction of RT initiatives, they were able to enjoy better livelihoods through engagement with the tourism industry.

Village life experience at Thekkady

Village life experience package is an innovative initiative which provides a better understanding of real-life situations and local culture in the local community. In this programme, economic benefits of tourism are equally distributed to the local community. Guests are accompanied by experienced tour guides to provide accurate information about the destination. Guests can interact with local people involved in each activity. Some of the activities in the village life experience package are traditional flower garland making, weaving of Panampu/Muram, fishnet stitching, bamboo curtain making, weaving of paya, and visits to spice-garden and bee-keeping units.

The study also revealed that local people have a strong desire to introduce their products to the tourism industry. The implementation of RT has helped them realize this aim as the focus is on local community development. The RT initiatives in Kerala are closely linked to community-based and rural tourism (Carlsen et al., 2008; Kokkranikal & Morrison, 2002; Mathew & Sreejesh, 2017; Thimm, 2016). By interlinking rural tourism initiatives with local people, RT was able to generate positive impacts such as foreign exchange earnings, employment generation, an enhanced standard of living and local community development. DLRTC also provided classes on making of cloth and paper bags and language training programmes to help locals develop greater confidence to interact with visitors. Among those who were interviewed, eight participants had attended some classes provided by the DLRTC. They also highlighted that in the initial stages the RT projects were carried out with financial assistance such as subsidy to startups and small-scale units. Hotels and resorts were initially reluctant to employ local people as they thought they would not have the required skills. With the introduction of this initiative, hoteliers and resorts' representatives interviewed mentioned that now they are providing more opportunities to local people for employment. So, it is evident that unemployment and poverty are steadily declining with the introduction of RT initiatives (Mathew & Sreejesh, 2017). Additional jobs have also been generated through several projects such as jeep safaris, taxi driving, homestays and tourist guides.

Socio-cultural impacts of responsible tourism in the local community

Kerala has different types of cultural art forms such as *Kathakali, Theyyam, Mohiniyattam and Thiruvathira*. The RT initiative helps to generate employment opportunities for cultural art performers in the destination. Some resorts and hotels arrange for a cultural night for visitors and employ local performers so as to generate income for the performers. As the local people lacked awareness of the negative social impacts of tourism, the Department of Tourism and the RT cell arranged awareness programmes on the adverse impacts of child labour, cultural inclusion and prostitution, to help the community understand the importance of tourism management in the destination by preventing such undesirable activities.

In the outskirts of Periyar Tiger reserve area, there live two types of tribes called Paliyan and Mannan. These tribes moved from Madurai in Tamil Nadu to the area in the 17th century. In the earlier decades, the tribal communities hesitated to interact with the non-tribal communities (Legrand et al., 2012). With the introduction of the India Eco-Development Project, the tribal communities obtained jobs in the Forest department within the Periyar Tiger Reserve. The RT cell also introduced a package to the tribal community to understand their culture, medicines and lifestyles. Some of the activities introduced in the RT initiatives focused on tribal food making, temple visits, tribal dances and demonstration and performance of *Paliyakoothu* musical instruments. In the initial years, the tribal communities supported these activities. After 2–3 years, they were reluctant to participate in the programmes because of privacy issues and fear of exploitation from others visiting their habitats. In addition, the Forest department also instructed the tribal community not to engage in non-tribal activities in their habitat, to prevent illegal activities such as poaching (Legrand *et al.*, 2012). The tribal dances are now performed in Vanasree auditorium in Thekkady every evening, for guests to experience and enjoy traditional art forms and culture.

Environmental impacts of responsible tourism in the local community

The RT initiatives help in the promotion of a clean environment with support for campaigns, cleaning programmes and meetings at the destination. A project called 'Clean Kumily Green Kumily' was organized by the Kumily Grama panchayath for maintaining cleanliness of the destination. This project has helped to create awareness among local people about the importance of cleanliness in the tourism industry. The 'Clean Kumily Green Kumily' project has employed around 40 local people to manage waste at the destination. They collect and segregate biodegradable waste, plastic and paper waste from the town, hotels, resorts and shops. Paper waste is treated and destroyed in different chambers to reduce carbon emissions. Biodegradable waste collected from hotels and the town is treated in different stages to make vermicompost for agriculture purposes. In the final stage, they segregate compost into different grades of vermicompost.

The study reveals that hotels and resorts have also been encouraged to engage in environmentally sustainable practices to achieve a platinum, gold or silver certificate with the introduction of RT initiatives. Some RT small-scale units are supplying cloth bags to hotels to avoid plastic waste. The RT initiative also focuses on a campaign called *free plastic zone* to reduce the use of plastic. In the interviews, resorts and hoteliers (3/3) were also encouraged to use renewable sources of energy such as solar panels for energy efficiency. Some hoteliers and resorts are also engaged in recycling water to conserve this important resource.

Tourism stakeholders and local community members highlighted water pollution problems during interviews. Sewage from the town and other liquid waste from different units go directly into the Periyar lake and affect the fish population in the lake. Local people also drew attention to chemical pesticides from cardamom plantations having some negative effects on the environment and water. In the earlier periods, normal wells were the major source of water for household needs and agricultural farming. Currently, people are facing water scarcity problems and most people in the locality are dependent on bore wells. As per local people's observation the ground water level is reducing every year. Some of the major findings and sustainability challenges provided by tourism stakeholders, employees and local community are outlined below.

- The scarcity of drinking water is one of the major concerns at the destination. The state government must ensure adequate steps are in place to harvest rain water in the destination to support farming and other needs.
- The government should entrust local bodies to function as watchdogs for controlling sewage waste into water bodies.
- The government must ensure and maintain the level of satisfaction among local people with regard to current tourism development and the potential expansion.

The identification of these findings and sustainability challenges provide a better opportunity for further research in green growth initiatives which will contribute to sustainable tourism development in the destination. The Indian state of Kerala does not qualify as a sustainable tourism destination, although there are individual success stories of non-governmental organizations (NGO) and at government level (Thimm, 2016). Green growth is an innovative concept to achieve effective sustainable or responsible tourism. In a developing country, green growth policies can reduce vulnerability to environmental risks and increase livelihood of the poor. Global Green Growth Institute, Korea has identified future challenges as greenhouse gas emissions, degradation of forests and water–energy–food shortages. The focus of green growth strategies is to ensure a practical and flexible approach for economic potential on a sustainable basis (Bowen & Fankhauser, 2011; Butler, 1999). That potential includes provision of critical life support services, clean air and water, and resilient biodiversity needed to support food production and human health. The latter include, for example, bringing more efficient infrastructure to people (e.g. in energy, water and transport), tackling poor health associated with environmental degradation and introducing efficient technologies that can reduce costs and increase productivity, while easing environmental pressure (Vazquez Brust & Sarkis, 2012). There is wide scope for expanding the green growth concept along with RT initiatives. The identification of these gaps provides a better opportunity for a further research agenda for green growth initiatives which will contribute to sustainable development.

Conclusion

In this chapter, we have examined the significant contribution of responsible rural tourism to destinations and how to interlink initiatives in different sectors with the local community in a developing country context.

It also reviewed the scope of the green growth approach to achieve sustainable development in the tourism destination context. The lack of research foci in green growth has limited the development of innovative measures for the conservation of resources. Knowledge of green growth initiatives has the potential to guide successful planning and improve and reorient strategies in RT, and enhance local community development in Kerala. This case study provides information about impacts of responsible rural tourism and its wider scope in a developing country context. This innovative initiative was able to generate positive linkages within the tourism industry, Department of Tourism and local community development. It also helped to explore site-specific activities and cultures through a village life experience package. Moreover, it helped to solve major issues in a tourist destination such as unemployment and poverty, to a certain extent.

Suffice to say that Kerala's RT initiatives have been able to showcase the region's unique selling proposition and tagline – Ayurveda, Backwaters, Culture and Cuisine (Chettiparamb & Kokkranikal, 2012). Through the Ayurveda industry at Kerala, the spice suppliers at Kimuli are able to enjoy the economic benefits of RT besides maintaining a unique cultural heritage that is distinctly Asian in its roots and philosophy. Ayurveda is an ancient Indian holistic healing treatment that focuses on the balance between the mind, body and spirit to promote good health instead of merely treating diseases. In essence Ayurveda embodies the Asian philosophy of balance and harmony based on the Indian cosmology. By leveraging on the thriving Ayurveda business in Kerala, the spice suppliers at Kimuli play a crucial role in supplying the essential oils and herbal ingredients that are vital for the Ayurveda process. In addition, the Kerala backwaters are the manifestation of balance between humans and nature, which has shaped the rural tourism imagery of the region.

References

Blackstock, K. (2005) A critical look at community based tourism. *Community Development Journal* 40 (1), 39–49.

Bowen, A. and Fankhauser, S. (2011) The green growth narrative: Paradigm shift or just spin? *Global Environmental Change* 21 (4), 1157–1159.

Brundtland, G.H. (1987) World commission on environment and development. *Environmental Policy and Law* 14 (1), 26–30.

Butler, R.W. (1999) Sustainable tourism: A 'state of the art' review. *Tourism Geographies* 1 (1), 7–25.

Cape Town Declaration (2002) Cape Town declaration on responsible tourism. In *Cape Town Conference on Responsible Tourism in Destinations*. See http://www.capetown.gov.za/en/tourism/Documents/Responsible%20Tourism/Toruism_RT_2002_Cape_Town_Declaration.pdfCarroll (accessed 22 March 2012).

Carlsen, J., Morrison, A. and Weber, P. (2008) Lifestyle oriented small tourism firms. *Tourism Recreation Research* 33 (3), 255–263.

Chettiparamb, A. and Kokkranikal, J. (2012) Responsible tourism and sustainability: The case of Kumarakom in Kerala, India. *Journal of Policy Research in Tourism, Leisure and Events* 4 (3), 302–326.

Clarke, J. (1997) A framework of approaches to sustainable tourism. *Journal of Sustainable Tourism* 5 (3), 224–233.
Debicka, O. and Oniszczuk-Jastrzabek, A. (2014) Responsible tourism in Poland. Paper presented at. *Biennial International Congress. Tourism and Hospitality Industry,* Faculty of Tourism and Hospitality Management, Opatija.
Eber, S. (ed.) 1992) Beyond the green horizon: A discussion paper on *Principles for Sustainable Tourism.* Godalming: Worldwide Fund for Nature.
Ebitu, E. (2010) Promoting an emerging tourism destination. *Global Journal of Management and Business Research* 10 (1), 21–24.
Edward, M. and George, B.P. (2008) Destination attractiveness of Kerala as an international tourist destination: An importance-performance analysis. *Proceedings of the Conference on Tourism in India – Challenges Ahead,* 15–17 May 2008, IIMK, Calicult, India.
Frey, N. and George, R. (2010) Responsible tourism management: The missing link between business owners' attitudes and behaviour in the Cape Town tourism industry. *Tourism Management* 31 (5), 621–628.
Goodwin, H. and Francis, J. (2003) Ethical and responsible tourism: Consumer trends in the UK. *Journal of Vacation Marketing* 9 (3), 271–284.
Harrison, L.C. and Husbands, W. (1996) *Practicing Responsible Tourism: International Case Studies in Tourism Planning, Policy and Development*: New York: John Wiley and Sons.
India Today (2017) Kerala's 'Responsible Tourism' initiative, a role model, says UNWTO. See https://www.indiatoday.in/travel/india/story/kerala-responsible-tourism-role-model-unwto-asia-india-lifetr-958972-2017-02-04 (accessed 3 January 2019).
Government of Kerala (2017) Kerala Tourism Statistics 2017. See https://www.keralatourism.org/tourismstatistics/tourist_statistics_2017_book20181221073646.pdf (accessed 26 May 2018).
Kokkranikal, J. and Morrison, A. (2002) Entrepreneurship and sustainable tourism: The houseboats of Kerala. *Tourism and Hospitality Research* 4 (1), 7–20.
Legrand, W., Simons-Kaufmann, C. and Sloan, P. (2012) *Sustainable Hospitality and Tourism as Motors for Development: Case Studies from Developing Regions of the World.* New York: Routledge.
Leslie, D. (2012) *Responsible Tourism. Concepts, Theory and Practice*: Wallingford: CABI.
Mathew, P. and Kumar, R. (2014) Responsible tourism-a grass root level empowerment mechanism: Case study from Kerala. *Innovative Issues and Approaches in Social Sciences* 7 (1), 53–73.
Mathew, P.V. and Sreejesh, S. (2017) Impact of responsible tourism on destination sustainability and quality of life of community in tourism destinations. *Journal of Hospitality and Tourism Management* 31, 83–89.
Okazaki, E. (2008) A community-based tourism model: Its conception and use. *Journal of Sustainable Tourism* 16 (5), 511–529.
Reed, M.G. (1997) Power relations and community-based tourism planning. *Annals of Tourism Research* 24 (3), 566–591.
Sharpley, R. and Telfer, D.J. (2014) *Tourism and Development: Concepts and Issues* (2nd edn). Bristol: Channel View Publications.
Spenceley, A. (2012) *Responsible Tourism: Critical Issues for Conservation and Development*: Abingdon: Routledge.
Thimm, T. (2016) The Kerala tourism model: An Indian State on the road to sustainable development. *Sustainable Development* 25 (1), 77–91.
UNEP (2010) *Green Economy Report: A Preview.* New York. See http://www.unep/greeneconomy/S (accessed 19 January 2018).
United Nations World Tourism Organization (UNWTO) (2007) WMO (2008) Climate Change and Tourism: Responding to Global challenges: Madrid: United Nations World Tourism Organization.

United Nations World Tourism Organization (UNWTO) (2011) Tourism towards 2030: Gyeongju, Republic of Korea: UNWTO.

United Nations World Tourism Organization (UNWTO) (2013) *Sustainable tourism for development guidebook: Enhancing capacities for Sustainable Tourism for development in developing countries*. Madrid: UNWTO Publications.

Vazquez Brust, D. and Sarkis, J. (2012) *Green Growth: Managing the Transition to a Sustainable Economy: Learning by Doing in East Asia and Europe*. Dordrecht; New York: Springer.

Venu, V. and Goodwin, H. (2008) The Kerala Declaration on responsible tourism. Paper presented at the *Incredible India, Second International Conference on Responsible Tourism in Destinations*, 21–24 March 2008, Kerala, India.

Vijayakumar, S.R.B.R. (2013) *Tourism and Livelihood*. Thiruvananthapuram: Kerala Institute of Tourism and Travel Studies (KITTS).

Theme 3
Tourism That Takes Place in Rural Areas

7 Linking Responsible Rural Tourism to Agritourism in the Philippines: A Case Study of Costales Nature Farms

Miguela M. Mena and Charmielyn C. Sy

Introduction

Reputed to be the world's largest industry and tagged as the largest employer, tourism has become a prominent element in the developmental strategies of many developing Asian nations (ADB, 2002). Due to this positive expectation in the growth of tourism, there is great optimism for rural tourism. Numerous agencies and academic researchers have identified tourism as a potential economic development tool, particularly for rural communities (Prosser, 2000; Wilkerson, 1996). Hence, tourism is increasingly being used to improve the social and economic well-being of residents in rural areas.

Rural tourism can be defined as the experience that takes place in agricultural or non-urban areas (Government of Alberta, Tourism, Parks & Recreation, 2010). Its essential characteristics include wide-open spaces, low levels of tourism development and opportunities for visitors to directly experience agricultural and/or natural environments. Rural tourism encompasses a huge range of activities, natural or manmade attractions, amenities and facilities, transportation, marketing and information systems (Sharpley & Sharpley, 1997). On the other hand, responsible rural tourism is defined as 'a quality travel experience that promotes conservation of the natural environment, protects the authenticity of culture, and offers socioeconomic opportunities and benefits for local communities'. This definition was adopted in 2002 at the Cape Town's World Summit on Sustainable Development.

The diversity of attractions within rural tourism includes heritage/culture-based, nature-based and agricultural areas. A major type of rural tourism is agritourism, which refers to visiting a working farm or any agricultural, horticultural or agribusiness operation for enjoyment, education, or active involvement in the activities of the farm or operation and includes taking part in a broad range of farm-based activities. Agritourism can be an example of responsible rural tourism and can bring economic advantages to farmers, community members and other stakeholders if it is developed responsibly and sustainably.

As an alternative strategy to agricultural development, agritourism responds to the need for responsible and sustainable rural development. From an economic aspect, farms involved in offering tourism experiences have managed to increase and diversify income streams and open employment opportunities to locals in their community (Barbieri, 2010; Che et al., 2005; Tew & Barbieri, 2012). It helps sustain small, family-operated farms while providing a venue for employment for other locals. In terms of its social sustainability, previous studies have also cited personal and societal benefits brought about by agritourism such as the promotion of farmer's identity as farming experts in educating visitors; conservation and preservation of the 'sense of place'; as well as cultivation of a 'sense of pride' for the locals in the area (Barbieri, 2010; McGehee & Kim, 2004; Ollenburg & Buckley, 2007; Tew & Barbieri, 2012). Farmers who can find additional income streams from tourism activities offered in their farms have managed to 'preserve their family heritage', which refers to the farms that have been handed down from one generation to another within the family (McGehee, 2007; Ollenburg & Buckley, 2007; Tew & Barbieri, 2012). From the guest's perspective, social impacts of agritourism are beneficial as well. Guests who visit the farms are provided with an opportunity to understand how the countryside impacts and supports the community and the economy (McGehee, 2007). Further, a visit to a farm provides the venue with a chance to rekindle ties with rural communities and get immersed in local customs and everyday rural living (Carpio et al., 2008; Veeck et al., 2006). Sonnino (2004: 285) cited agritourism as a responsible and sustainable strategy, such that 'it promotes the conservation of a broadly conceived rural environment through its socioeconomic development'. Accordingly, she further stated that agritourism improves both the natural and built environments within farm areas and enhances the environmental conservation and management of the resource it builds upon (Sonnino, 2004). Agritourism encourages the operations of low-impact agricultural and tourist activities. Hence, it promotes responsible, ethical and sustainable rural tourism that benefits local communities who primarily depend on agriculture.

Agritourism, as a special form of tourism, links resources, products and services from both the agricultural and tourism sectors that would provide new experiences. The linking of tourism activities with farm resources has gained considerable attention because of its potential benefits to farmers and rural communities. Literature concentrates on its socio-economic value with the term encompassing all activities inside a working farm for educational or leisure purposes of the visitors and generates additional income streams for the farmer or farm owners (Barbieri & Mahoney, 2009; Ollenburg & Buckley, 2007; Tew & Barbieri, 2012; Veeck *et al.*, 2006). Some research distinguishes agritourism as a part of the bigger concept of rural tourism by including other tourism components such as accommodation, attractions and amenities in rural areas. (Rogerson & Rogerson, 2014).

The most common benefit that agritourism generates for farms is in generating new revenue streams (Bwana *et al.*, 2015; Flanigan *et al.*, 2014, 2015; Naidoo & Sharpley, 2016). Other economic benefits include making full use of farm resources, mitigating farm product price instability and providing additional employment for other family members and the local community (McGehee *et al.*, 2007; Schilling & Sullivan, 2014; Schilling *et al.*, 2012). Social and non-quantifiable benefits include inspiring aesthetic and cultural values of the place, sustaining the responsible use of the land resource and uplifting the farmer's identity (Bwana *et al.*, 2015; Flanigan *et al.*, 2015; Naidoo & Sharpley, 2016). The chances to experience rural life and be educated about farming are also some of the identified benefits of agritourism from the visitors' perspective (Busby & Rendelle, 1999; Flanigan *et al.*, 2014). Activities such as educational and leisure farm tours, fruit picking, fishing, nature and bird watching, as well as recreational harvesting, preparing and consuming farm produce, seminars and workshops on farming and holding of events and shows are included in the portfolio of agritourism products (Barbieri *et al.*, 2008, Tew & Barbieri, 2012). The motivation to transition from purely agricultural pursuits to agritourism is driven primarily by the perceived benefits of the latter (Barbieri, 2010; McGehee *et al.*, 2007; Tew & Barbieri, 2012).

Transformative tourism is a developing concept in the discussion of sustainability. Reisinger (2013a) opined that the transformative value of agritourism, or farm tourism, lies in the involvement of guests in the rural activities and the subsequent learning and understanding they achieve from a tourism experience. The transformative value, from the hosts' perspective on the other hand, is the realization of the perceived benefits that agritourism offers which motivated them to engage in the activity in the first place (Reisinger, 2013b).

This chapter explores agritourism as a responsible rural tourism product in the Philippines by examining its first accredited agritourism site, the Costales Nature Farms. The chapter is organized as follows: firstly, a discussion of how agritourism developed in the Philippines is presented; this is followed by a general description of the different types of agritourism sites in Philippines and the case of Costales Nature Farms, the first accredited farm tourism site in the Philippines; and thirdly, the conclusion of the chapter is presented together with the challenges in developing agritourism.

Methodology

An extensive literature review on 'sustainable tourism', 'responsible tourism', 'rural tourism', 'responsible rural tourism', 'agritourism' and 'rural development' was undertaken to become familiar with the terms, definitions, key organizations and stakeholders involved, and tourism types and approaches (e.g. pro-poor tourism, responsible tourism, ecotourism). To accomplish the objectives of the chapter, a case-study methodology is used in conjunction with secondary data analysis and key informant interviews.

Linking agriculture and tourism: The development of Philippine agritourism

Agriculture and tourism are the 'two most dynamic industries in the Philippines' (Icamina, 2012). Both industries generate employment (Cayabyab, 2013) and have the capability to benefit all regions and provinces of the country (Marcos Jr., 2009). Agriculture and tourism represent a 'perfect tandem' complementing each other, since agriculture can be packaged as a 'tourism product' and developed for tourism consumption (Lesaca, 2012). Agritourism, the combination of tourism and agriculture (Freznosa, 2012), is defined as 'a commercial enterprise at a working farm, ranch or agricultural plant conducted for the enjoyment of the visitors that generates supplemental income for the owner' (University of California Small Farm Program, 2011). In the Philippines, the Department of Tourism (DOT) defines agritourism as 'a form of tourism activity conducted in a rural farm area which may include tending to farm animals, planting, harvesting and processing of farm products. It covers attractions, activities, services and amenities as well as other resources of the area to promote an appreciation of the local culture, heritage and traditions through personal contact with local people'.

Agritourism is also defined as a 'subset of a larger industry called rural tourism' (Bernardo *et al.*, n.d.) which encompasses exposure to and contact with residents, customs, culture, way of life and activities in the surroundings (Wolfe & Bullen, n.d.) It is considered as 'an activity, enterprise or business that combines primary elements and characteristics of agriculture and tourism and provides an experience for visitors that stimulates economic activity and impacts both farm and community income' (Wolfe & Bullen, n.d.). It is synonymous to and used interchangeably as 'agricultural tourism, agri-tainment, agro-tourism, farm tourism, [and] farm visits' (Rich *et al.*, 2005).

In 1991, the DOT and the United Nations Development Programme (UNDP) formulated the Philippine Tourism Master Plan (TMP) which aimed to develop tourism that is environmentally sustainable (Spire Research and Consulting, 2013). The process of transforming agricultural and fishing sites into tourism attractions has been present in the Philippines since the 1990s but was not institutionally defined (Spire Research and Consulting, 2013). Agritourism sites were initially identified and recognized by both the Department of Agriculture (DA) and DOT in 1999. Both national government agencies envisioned a collaborative industry that would cascade the benefits derived from tourism to the agricultural sector in the rural areas. An agritourism programme was conceptualized and formed through the collaboration of DOT, DA and the University of the Philippines Asian Institute of Tourism (Cruz, 2003) and was launched in the latter part of 2002 with the goal of making the country a premier agritourism destination in Asia.

In 2002, the DA and DOT worked with the University of the Philippines Asian Institute of Tourism (UP AIT) on a manual that identified agritourism sites in the country (SEARCA, 2011). Initial agritourism sites in the Philippines such as C & B Orchid Farm in San Rafael, Bulacan, Sonya's Secret Garden in Alfonso, Cavite, Oroverde in Guimaras and Del Monte Plantation in Manolo Fortich, Bukidnon were identified (SEARCA, 2011).

In 2012, the House of Representatives filed the House Bill 1808 to promote agritourism in the country through (a) tax credits for registered activities to help offset the expenses of those venturing into agritourism and (b) technical assistance to farmers entering the agritourism business.

In 2016, the Republic Act 10816 was enacted providing for the development and promotion of farm tourism in the Philippines. The law lays down the functions and the organizational structure of the newly created agency, the Philippine Farm Tourism Industry Development Coordinating Council (PFTIDCC), which will be administratively attached to the DOT. The tasks of the PFTIDCC are to (a) prepare and implement a Comprehensive National Farm Tourism Industry Development Plan; (b) establish and maintain a comprehensive farm tourism information system; (c) formulate farm tourism research and development projects; and (d) provide for the registration and accreditation of farm tourism practitioners and operators.

Aside from the DA and the DOT, other agencies such as the Department of Environment and Natural Resources (DENR), Department of Education (DepEd), Commission on Higher Education (CHED), Department of Science and Technology (DOST), Department of Public Works and Highways (DPWH), Department of Trade and Industry (DTI), Philippine Information Agency (PIA), Department of Local and Interior Government (DILG), National Economic Development Authority (NEDA) and local government units have been tasked to work together for the efficient operations of the PFTIDCC.

The DOT and the DA, in collaboration with the other agencies mentioned, will formulate a six-year farm tourism strategic development plan which will be evaluated and updated every three years. The development plan includes the (a) identification of farm tourism sites; education and promotional support; (b) infrastructure, investment and market promotion; (c) possible extension programmes; and (d) research and development. The PFTIDCC is mandated to encourage the establishment of at least one tourism farm in every province in the country. Accreditation of the PFTIDCC shall be valid for two years only. However, due to a change in the administration following the May 2016 national elections, the drafting of the implementing rules and regulations of the law has been put on hold.

Categories of the agritourism sites in the Philippines

There are two categories of agritourism sites in the Philippines: the day farm, and the farm resort (refer to Table 7.1 for selected agritourism sites in the Philippines). Day farms are those located near national highways and main business areas and are ideal for day tours and visits. On the other hand, farm resorts offer accommodations and dining services. Both hold interactive on-farm activities and other attractions that can

Table 7.1 Selected farm tourism sites in the Philippines accredited by the DA or DOT

Name of farm	Location (Province)	Category	
		Day farm	Farm resort
Gourmet Farm	Cavite	√	
Sonya's Garden	Cavite	√	√
Mindanao Baptist Rural Life Center	Davao	√	√
Trappist Monastery	Guimaras	√	
Ilog Maria Honeybee Farm	Cavite	√	
Bohol Bee Farm Resort / Restaurant	Bohol	√	√
Del Monte Philippines Inc.	Bukidnon	√	
MenziAgri Development	Bukidnon	√	
Tomato Farms	Ilocos Norte	√	
National Agriculture Research	Ilocos Norte	√	√
Central Luzon State University	Nueva Ecija	√	√
University of the Philippines Los Banos	Laguna	√	√
Los Banos Horticulture Society	Laguna	√	
Hacienda Macalauan Inc	Laguna	√	
Pamora Farm	Abra	√	
Costales Nature Farm	Laguna	√	√
Palawan Butterfly Garden	Palawan	√	
Joboken Farm Enterprise	Albay	√	
Gawad Kalinga Enchanted Farm	Bulacan	√	√
Duran Farms	Bulacan	√	√

Sources: Esplana, 2011; farm visits and websites.

enrich tourists' experience in farm life and are generally located at safe and peaceful areas.

Both day farms and farm resorts are required to have a reception counter, parking space, a dining/multi-purpose area, farm guide(s) and a souvenir shop. For the farm resort, two additional elements required are accommodation and a restaurant. Accreditation is through a DOT-issued certification officially recognizing the site as having complied with the minimum standards and requirements prescribed for the operation and maintenance of the agritourism farm.

Agritourism destinations that the DOT accredits are featured in all the promotional programmes that the department implements. They also benefit from being brought into the extensive network of local contacts built up by the DOT over the years.

Currently there are several popular agritourism sites. The Gawad Kalinga Enchanted Farm considers itself as a 'farm village university' where communities live together, study technical courses and do farming while allowing visitors to experience the life on the farm through different activities such as hiking, bird-watching, campfires, 'harvest-your-own' fruit and overnight stays. The Duran Farm is where local and foreign visitors can learn best practices in vegetable production and serves as a venue for private companies conducting seminars and training programmes. Other farms are the Leisure farms in Batangas, the Mango farms in Guimaras, Bohol Bee farm in Bohol, Dragon Fruit farm in Ilocos Norte and the Rice farms in Negros Occidental. In some provinces, agritourism is practised by leasing land for a period so that tourists can grow and harvest their own produce. The strawberry and organic vegetable farms of Benguet are also well-known for the pick-your-own-fruit activities. Bukidnon, which has vast pineapple and coffee plantations, is tagged as one of the ideal spots for agritourism.

The Philippines has accredited more than 30 agritourism sites since 2012 ranging from huge pineapple plantations to small orchid farms, bee farms, or plantations that specialize in growing exotic tropical fruits such as dragon fruit and papaya. It also includes 27 protected areas like the Tubbataha Reef, a national park in Palawan. The Philippines also hosts various events, including farmers' field days and agricultural fairs, which are estimated to be attended by 64,000 tourists each year.

Case Study: Costales Nature Farms

Costales Nature Farms is the country's first accredited agritourism destination by the Department of Tourism (Costales Nature Farms, 2015). This 7700 sq m farm is in Barangay Gagalot, Majayjay Laguna. It is an integrated and commercial farm that practices sustainable and organic farming. The farm functions both as farm producer and experience provider.

In 2005, Ronald and Josephine Costales decided to develop a plot of hilly land as a weekend getaway from their home in Sta. Rosa City, as well as a sustainable source of organic vegetables, fish and livestock for the family table. In 2008, the couple ventured into the commercial growing of organic vegetables and fish and gradually expanded their farm area. Eventually, they became the country's biggest single producer of organically grown, high-value vegetables and culinary herbs.

By 2012, Costales Nature Farms' total land area reached five hectares through acquisition of neighbouring lands. It was also certified as an organic producer by the Negros Island Certification Services Inc. (NICERT), one of the country's two DA-accredited organic certification bodies for Internal Control System groups and individual farmers nationwide. An organic certification is a way of assuring that the products consumers buy is safe from harmful chemicals. The farm then worked for accreditation with the DOT as a fully fledged agritourism destination, the very first in the country.

The farm was not hard to promote since it is in the proximity of Metro Manila and has a climate that remains mild even in summer, providing natural 'air-conditioning' for accommodation. It is a diversified integrated farm with a variety of organic and healthy products. Visiting and seeing the farm first hand assures people of where its products come from and that these are really organic and grown only with organic fertilizer produced on the farm itself through vermiculture and composting of plant and animal wastes.

The designation as the official agritourism destination for an international beauty pageant (2013 Ms. World Philippines) further boosted the farms' popularity. Since then, the farm has been hosting 3000 to 4000 visitors a month, 95% of whom are local tourists, and slightly over 10% stay overnight. The farm can accommodate 60 people who wish to stay for more than a day. The lodging facilities range in sizes from shaw-roofed and intricately built bamboo native huts for two persons to dwellings for families with up to 10 members.

Visitors can choose from five modes of stay. Its 'Wellness Tour' is for three days and two nights of full rural relaxation and farm life experience in a healthy and natural environment, while the 'Life at the Farm Tour' is a two-days-and-one-night stay. The 'Green Living Tour' is a whole day of relaxation with an eco-tour, and the 'Green Salad Tour' is a half-day experience of farm life. The 'Lakbay-Aral Tour' is especially designed for students and farmers to impart and spread healthy farming to encourage future generations appreciate organic food and a farm lifestyle.

All the tours involve an orientation on organic farming, a guided farm tour and enjoying vegetable salad snacks. The first three packages include lunch, unlimited salad snacks and fishing activities. The Wellness and Life-at-the-Farm visitors have personal vegetable harvesting for breakfast and lunch, while the same activity is on a pick-and-pay basis for the other tours, with demonstrations on salad preparation.

The farm's business partners include the country's largest consolidator of fresh fruits and vegetables that distributes the farm products to supermarket chains. Another is the number one wellness store chain in the country, which exclusively sells organic free-range chicken meat, eggs, pork and vegetables, which help supply the farm with top-of-the line, post-harvest facilities to ensure the highest standards of food hygiene. Another partner is the biggest chain of fine high-end restaurants and dining concepts to produce high-value vegetables and culinary herbs that are served with the label 'I Love Organic'.

The owners of the farm have also forged tie-ups with the DA's Agricultural Training Institute and other government agencies and private foundations for promoting organic agriculture. The Technical Education and Skills Development Authority (TESDA) also accredits their facility to train, test, and certify organic farm workers.

The farm is a perfect model of not only certified organic production but also sustainability, income diversification, business partnership, agritourism, extension service, creative fund generation and the management of interaction with people. It is also a perfect model of responsible rural tourism. The farm was awarded the 2015 Model Agritourism destination and the First DOT accredited agritourism in the Philippines and was officially tapped by DOT on 9 February 2016 as their training and extension partner in promoting the development of agritourism throughout the country.

Conclusion

Diverting agricultural lands into agritourism is not an easy process. Stakeholders face several challenges before they are able to reap the success of diverting their farms into agritourism. As seen in the case of Costales Nature Farms, agritourism development requires substantial investment to provide standard facilities and to meet the costs of the development (Ambarawati, 2014). The location of an agritourism destination can also hinder its development. Most agritourists travel only a short distance to agritourism destinations. Farms located near populated areas clearly have location advantage. The lack of research, planning or informed decision making is also a constraint in agritourism development (Ryan *et al.*, 2006). The seasonality characteristic of agritourism and it being dependent on weather conditions have also been identified as one of the difficulties faced in agritourism development (Ryan *et al.*, 2006).

Based on the experience of Costales Nature Farms, the success of agritourism development seems to depend on the following features: specific legislation or a regulatory framework for agritourism, extensive education and training support for agritourism enterprises and marketing support. Regarding the specific legislation or a regulatory framework for agritourism, the Philippines already has this with the accreditation of day farms and farm resorts, but it is limited, overlooking most of the more popular agritourism activities. The accreditation or certification earned for the enterprise is a prerequisite for marketing, business development and financial support. However, the implementing rules and regulations of the Farm Tourism Act are still being formulated. For the second feature, extensive education and training support for agritourism enterprises, DA and DOT have started to address this aspect. They recognize that the enterprises are farm businesses first, and not necessarily well versed in the tourism and hospitality industry. Educating farming enterprises about the opportunities for supplemental (or even primary) revenue from agritourism helps to grow the industry and improve the economic state of farming communities and is critical for the support of new agritourism enterprises to ensure their success. The third feature is marketing support. The general pattern of marketing efforts in successful programmes abroad is for government tourism agencies to handle the marketing (internet, print and television) on a national or an international scale, and provide some level of advertising subsidy to individual accredited agritourism enterprises or cooperatives to offset costs of their own marketing efforts. Local marketing can be handled at a community level. The national and local government units should collaborate in marketing accredited agritourism sites.

Tourism activities are considered as sustainable only when they are economically viable without destroying the environment and when the social wellbeing of the community is upgraded (Swarbrooke, 1999). In the case of Costales Nature Farms, sustainable principles are translated into the ability to preserve the farmer's historic ties to the land and traditional knowledge, employing sustainable agricultural practices, increasing farm revenues and profits, and sustaining landscape, habitats and soil productivity. Consequently, it is possible to preserve the family farmland for future generations and sustain rural economies. Agritourism produces multiple environmental, sociocultural and economic benefits for their farms, households and even society. It increases the wellbeing of rural families, contributing to the employment of family members and future generations, and for non-family members, attracting the youths to live in rural communities. Agritourism activities strengthen the preservation of the rural landscape and sensitize customers to preserve nature and to safeguard the environment. Costales Nature Farms shows that agritourism can open new paths toward rural sustainable development and responsible rural tourism.

However, having an institutionalized system of accrediting agritourism farms might lead to commodification and the lack of differentiation. While the current activities and experiences offered by the certified farms might be able to meet the needs of predominantly urban dwellers, their content and presentation are mostly designed by related government agencies. The relative lack of involvement by the farming enterprises in the planning process poses three critical challenges for the sustainability of this form of institutionalized agritourism: (1) How to ensure that the farming enterprises fully understand the demand for agritourism and match the tourist expectations with creative experiences based on their farms' unique selling point. (2) How to showcase the farm experience beyond specific activities for urban families to a holistic interpretation of the socio-ecological processes of the rural landscape in the Philippines. Obviously it would not be possible for agritourism in the Philippines to replicate the majesty of the Banaue rice terraces, which is the first cultural landscape to be listed in United Nations Educational, Scientific and Cultural Organization (UNESCO)'s World Heritage List. However, the underlying philosophy of balance between humans and nature that is manifested in the Banaue rice terraces is also evident in many of the traditional farms in the Philippines. Adding the Asian philosophy could significantly add value to the experience of demonstration farms such as Costales Nature Farms, beyond showcasing the historic ties to the land and traditional knowledge. (3) The final challenge is how to enhance the environmental education dimension of the agritourism farms. As development pressure is threatening rural land uses, the agritourism farms should able also add depth to the story telling by highlighting the contribution of sustainable landscape management towards nature-based solutions in biodiversity conservation, which is essentially Asian in its philosophy.

References

Ambarawati, B.I.W. (2014) Community based agro-tourism as an innovative integrated farming system development model towards sustainable agriculture and tourism in Bali, Indonesia. *Journal of the International Society for Southeast Asian Agricultural Sciences* 20 (1), 29–40.

Asian Development Bank(ADB) (2002) *Asian Development Outlook 2002*. New York: Oxford University Press

Austria, J.D. and Horigue, M.A. (2014) A Philippine agritourism systems model: A Weberian approach of the case of Majayjay, Laguna. Unpublished thesis. Asian Institute of Tourism, University of the Philippines, Diliman, Quezon City.

Barbieri, C. (2010) An importance-performance analysis of the motivations behind agritourism and other farm enterprise developments in Canada. *Journal of Rural and Community Development* 5 (1), 1–20.

Barbieri, C. and Mahoney, E. (2009) Why is diversification an attractive farm adjustment strategy? Insights from Texas farmers and ranchers. *Journal of Rural Studies* 25 (1), 58–66.

Barbieri, C., Mahoney, E. and Butler, L. (2008) Understanding the nature and extent of farm and ranch diversification in North America. *Rural Sociology* 73 (2), 205–229.

Bernardo, D., Leatherman, J. and Valentin, L. (n.d.) Agritourism: If we build it, will they come? Agritourism. University of California Cooperative Extension. See http://sfp.ucdavis.edu/agritourism/ (accessed 27 July 2018).

Bwana, M.A., Olima, W.H., Andika, D. Agong, S. and Hayombe, P. (2015) Agritourism: Potential socio-economic impacts in Kisumu county. *Journal of Humanities and Social Sciences* 7 (3), 78–88.

Busby, G. and Rendle, S. (1999) Transition from tourism on farms to farm tourism. *Tourism Management* 21 (6), 635–642.

Carpio, C.E., Wohlegenant, M.K. and Boonsaeng, T. (2008) The demand for agritourism in the United States. *Journal of Agricultural and Resource Economics* 2 (33), 254–269.

Cayabyab, M. (2013, 1 May) Agriculture: The decline of the poor man's sector. *Economy |GMA News Online*. See http://www.gmanetwork.com/news/story/306370/economy/agricultureandmining/agriculture-the-decline-of-the-poor-man-s-sector (accessed 19 May 2017).

Che, D., Veeck, A. and Veeck, G. (2005) Sustaining production and strengthening the agritourism product: Linkages among Michigan agritourism destinations. *Agriculture and Human Values* 22 (2), 225–234.

Costales Nature Farms (2015) About us. See http://www.costalesnaturefarms.com (accessed 11 August 2018).

Cruz, R.G. (2003) Towards sustainable tourism development in the Philippines and other ASEAN countries: An examination of programs and practices of National Tourism Organizations. *PASCN Discussion Paper*, No. 2003–06

Esplana, E.R. (2011) Development in the supply chain of the Philippine agritourism industry: An assessment. See http://www.slideshare.net/cpr_elmer/development-in-the-supply-chain-of-the-philippine-agritourism-industry (accessed 11 August 2017).

Flanigan, S., Blackstock, K. and Hunter, C. (2015) Generating public and private benefits through understanding what drives different types of agritourism. *Journal of Rural Studies* 4, 129–141.

Flanigan, S., Blackstock, K. and Hunter C. (2014) Agritourism from the perspective of providers and visitors: A typology-based study. *Tourism Management* 40, 394–405.

Freznosa, E.P. (2012) Harnessing agritourism opportunities in the Philippines. See https://prezi.com/3wcertsiqh6s/harnessing-agritourism-opportunities-in-the-philippines/ (accessed 6 December 2013).

Government of Alberta, Tourism, Parks and Recreation (2010) Tourism works for Alberta: The economic impact of tourism in Alberta. See http://www.tpr.alberta.ca/tourism (accessed 28 June 2016).

Icamina, P. (2012) Visitors willing to pay for agritourism, conservation See http://ovcre.uplb.edu.ph/index.php?option=com_k2&view=item&id=129:visitors-willing-to-pay-for-agritourism-conservation&fb_ref=Default,@Total (accessed 14 May 2015).

Lesaca, P. (2012) Agriculture and Tourism: The Perfect Tandem. BAR Digest Home, 14(3). See http://www.bar.gov.ph/digest-home/digest-archives/367-2012-3rd-quarter/4423-julsep2012-agriculture-tourism (accessed 30 September 2017).

Marcos, Jr., F. (2009) BongBongMarcos.com. See http://www.bongbongmarcos.com/news/post/the-importance-of-infrastructure-and-tourism-development (accessed 2 February 2009).

McGehee, N.G. (2007) An agritourism systems model: A Weberian perspective. *Journal of Sustainable Tourism* 15 (2), 111–124.

McGehee, N.G. and Kim, K. (2004) Motivation for agritourism entrepreneurship. *Journal of Travel Research* 43, 161–170.

McGehee, N.G., Kim, K. and Jennings, G.R. (2007) Gender and motivation for agritourism entrepreneurship. *Tourism Management* 28, 280–289.

Naidoo, P. and Sharpley, R. (2016) Local perceptions of the relative contributions of enclave tourism and agritourism to community well-being: The case of Mauritius. *Journal of Destination Marketing and Management* 5 (1), 16–25. https://doi.org/10.1016/j.jdmm.2015.11.002

Ollenburg, C. and Buckley, R. (2007) Stated economic and social motivations of farm tourism operators. *Journal of Travel Research* 45 (4), 444–452.

Prosser, G. (2000) Regional tourism research: A scooping study. Occasional Paper. Number 4. Lismore: Southern Cross University.

Rich, S., Barbieri, C. and Arroyo, C. (2005) Agritourism. University of California Cooperative Extension. See http://sfp.ucdavis.edu/agritourism/ (accessed 13 March 2018).

Reisinger, Y. (2013a) Transformation and transformational learning theory. In Y. Reisinger (ed.) *Transformative Tourism: Tourist Perspectives* (pp. 17–26). Wallingford: CABI.

Reisinger, Y. (2013b) Connection between travel, tourism and transformation. In Y. Reisinger (ed.) *Transformative Tourism: Host Perspectives* (pp. 27–31) Wallingford: CABI.

Rogerson, C.M. and Rogerson, J.M. (2014) Agritourism and local economic development in South Africa. *Bulletin of Geography, Socio-Economic Series* 26, 93–116.

Ryan, S., Debord, K. and McClellan, K. (2006) Agritourism in Pennsylvania: An industry assessment. See http://www.rural.palegislature.us/documents/reports/agritourism2006.pdf (accessed 12 August 2018).

Schilling, J., Opiyo, F.E.O and Scheffran, J. (2012) Raiding pastoral livelihoods: Motives and effects of violent conflict in north-western Kenya. *Pastoralism: Research, Policy and Practice* 2 (25), 1–16.

Schilling, B.J. and Sullivan, K.P. (2014) Characteristics of New Jersey agritourism. *Journal of Food Distribution Research* 45 (2), 161–173.

Swarbrooke, J. (1999) *Sustainable Tourism Management*. New York: CABI Publishing.

Southeast Asian Regional Center for Graduate Study and Research in Agriculture (SEARCA) (2011) Agritourism in the Philippines – untapped potential. See https://www.searca.org/events/seminar/2011/agritourism-in-the-philippines-untapped-potential (accessed 27 June 2018).

Sharpley, R. and Sharpley, J. (1997) *Rural Tourism: An Introduction*. London: International Thomson Business Press.

Spire Research and Consulting (2013) The rise of agri-tourism in the Philippines. See https://www.spireresearch.com/spire-journal/yr2013/q3/the-rise-of-agri-tourism-in-the-philippines/ (accessed 3 June 2018).

Sonnino, R. (2004) For a 'piece of bread'? Interpreting sustainable development through agritourism in Southern Tuscany. *Sociologia Ruralis* 44 (3), 285–300.

Tew, C. and Barbieri, C. (2012) The perceived benefits of agritourism: The provider's perspective. *Tourism Management* 33, 215–224.

University of California Small Farm Program (2011) What is Agritourism? See http://sfp.ucdavis.edu/agritourism/factsheets/what/ (accessed 11 March 2017).

Veeck, G., Che, D. and Veeck, A. (2006) America's changing farmscape: A study of agricultural tourism in Michigan. *The Professional Geographer* 58 (3), 235–248.

Wilkerson M.L. (1996) Developing a rural tourism plan: The major publications. *Economic Development Review* 14 (2), 79.

Wolfe, K. and Bullen, G. (n.d.) Considering an Agritourism Enterprise? See http://ag-econ.ncsu.edu/sites/agecon.ncsu.edu/files/faculty/bullen/Considering%20an%20Agritourism%20Enterprise.pdf (accessed 15 July 2017).

8 Is Community-Based Tourism a Tool for the Sustainability of the Local Community and the Local Economy? The Case of Coruh Valley, Turkey

Sıla Karacaoğlu and Medet Yolal

Introduction

The United Nations Development Programme (UNDP) has launched a bundle of projects to increase capacity usage for better democratic governance to eradicate poverty and reduce inequalities through sustainable development, and improve sustainability of environment and development in Turkey, along with several other countries around the globe. As suggested by Lapeyre (2010), building on new political paradigms of people's participation, the ownership and operation of tourism ventures by indigenous people themselves is now increasingly seen as one of the seven mechanisms by which tourism could efficiently help reduce poverty. In this regard, a community participation approach has long been advocated as an integral part of sustainable tourism development (Okazaki, 2008).

Sustainable programmes and projects focusing on community-based tourism (CBT) are increasingly initiated in less developed and underdeveloped countries. According to Hiwasaki (2006), there are four objectives of CBT: (1) to increase local community empowerment and ownership through participation in the planning and management of tourism in protected areas; (2) to have a positive impact on conservation of natural and or cultural resources in and around protected areas through tourism; (3) to enhance or maintain economic and social activities in and around protected areas, with benefits (economic and social) to the local community; and (4) to ensure that visitor experience is of high quality and is socially and environmentally responsible. Nevertheless, there are researchers like Blackstock (2005) who argues that CBT is an unrealistic discourse for legitimizing tourism development. They claim that CBT sidesteps the issues of social justice and local empowerment. Likewise, there is a tendency to accept CBT as an alternative to traditional economic activities. In fact, it should be seen as an activity which is complementary to, and never a substitute for, traditional activities based primarily on agriculture, fishing, and livestock farming (López-Guzmán et al., 2011). As such, Sebele (2010) suggests that unless local residents are empowered and participate fully in decision-making and ownership of tourism developments, tourism will not reflect their values and will be less likely to generate sustainable outcomes. Despite the critiques, Kibicho (2008) notes that CBT is an effective way of implementing policy coordination, avoiding conflicts between different actors in tourism, and obtaining synergies based on the exchange of knowledge, analysis and ability among all members of the community. However, there is evidence that the majority of CBT initiatives enjoy success (Lapeyre, 2010). CBT initiatives are generally small-scale, and it is not possible for all members of the community to be involved and thus derive benefits. Furthermore, CBT projects generally result in uneven distribution of the benefits of tourism development among the locals. Overall, Salazar (2012) argues that while CBT is intended to empower people, the representations deployed in constituting the targeted 'communities', be they imagined or real, remain largely unexamined. In this regard, Goodwin and Santilli (2009) note that although many projects have been funded in developing countries, their achievement has not been widely monitored and, therefore, the actual benefits to local communities remain largely unquantified.

One of these projects was the Eastern Anatolia Tourism Development Project initiated in Coruh Valley, Turkey in March 2007, under the partnership of UNDP, the Ministry of Culture and Tourism and a private Turkish company, Efes Pilsen. This chapter investigates the local business owners' perceptions and attitudes towards tourism development in their region. Data were collected from 46 small business owners in the region through telephone interviews and the respondents were asked to express their experiences, attitudes and perceptions of tourism before and after the project.

The Project

Eastern Anatolia Tourism Development Project

Coruh Valley is in the eastern part of the country, denoting the valley of the Coruh River. The region covers Tortum, İspir and Uzundere counties of Erzurum Province and Yusufeli county of Artvin Province. Eastern Anatolia Tourism Development Project aimed to improve the economic well-being of the local people through the development of income-generating industries as an alternative to agriculture in Coruh Valley (Kithir, 2012: 3). One of the alternative strategies to achieve the goals was the development of tourism, specifically alternative tourism. The project was initiated in 2007. The first phase of the project was completed in 2009, while the second phase was completed in 2012. The project was administered under the partnership of UNDP, the Ministry of Culture and Tourism and a private Turkish company, Efes Pilsen. The mission statement of the project was 'to develop a functional and participatory model in order to sustain natural and cultural resources of Coruh Valley in terms of tourism development'.

The status of the region prior to the project was documented in the project's website (Coruh, 2018). It was noted that while the region's natural and cultural resources were remarkable, the economic income generating opportunities were scarce. The people in the region were conservative and consequently female participation in the economic and social activities was limited. Moreover, people were not aware of the value of their natural, cultural and historical resources. Meanwhile, lack of infrastructure and superstructure hindered tourism development. There were no accommodation establishments in Uzundere and İspir counties. Boarding houses were in their infancy in Yusufeli. The number of service businesses such as restaurants, camping or picnic areas was also limited, offering low-quality services. Although local authorities were willing to develop tourism in their region, they had no expertise in this regard. Further, development was assumed to be a physical process and social and human aspects were ignored.

First phase (2007–2009)

The purpose of the project's first phase was threefold: to develop tourism products, to improve local capacity and to promote the region and the project. Developing tourism products required efforts such as making resources inventory, determining tour routes, supplying logistics support for travel agencies, organizing a birdwatching festival and preparing maps, brochures and leaflets about the region. Training local people was crucial during this phase. As such, more than 150 people were trained on the botanical structure of the valley. In order to promote entrepreneurship in the region, familiarization tours to developed regions were organized. Consequently, new boarding houses were opened by local people, most of whom were females. Alternative forms of tourism in the valley were developed in the course of time during the project. Bird, butterfly and bear watching tours, trekking, canoeing and rafting tours were organized (UNDP, 2018). Tour guides for trekking and mountain biking were trained and employed. Improving local capacity focused mainly on increasing the bed supply, training the local people and offering incentives to tourism entrepreneurs. Moreover, non-governmental organizations were empowered. In order to promote the region, signboards about the attractions were prepared and placed. Further, multilingual materials were distributed to potential tourists. A documentary film was prepared about the valley and its attractions. Cooperation with national tourism bodies for promotional purposes was triggered. The valley and the project were also promoted in industry fairs and meetings (DATUR Final Presentation, 2011).

Second phase (2010–2012)

The second phase of the project relied heavily on institutionalization. Therefore, a series of activities such as rehabilitating the accommodation businesses, female participation (founding associations, their inclusion into the workforce, empowering income generating activities), youth participation, attracting new funds to the region, sharing experiences, business development, training and capacity improvement were implemented.

Accommodation businesses were trained and supported on a continuous basis. Potential investors were encouraged to convert out-of-use village schools to boarding houses. In addition to improvements in the current bed supply, two new boarding houses were opened in the downtown of Uzundere with a capacity of seven rooms and 14 beds. In the Yedigöl region of Uzundere, a small motel with 24 beds was opened in 2011. Similar investments were made in several centres, providing technical and administrative support in the context of the project. Consequently, the total number of 26 businesses in 2007 increased to 51 by the end of 2010.

In terms of educational and training efforts, training courses were organized to increase the number of young people engaged in tourism activities. As such, local people received training in rural tourism, nature-based sports, communication skills, protection of natural and cultural resources and sustainability issues. The goal was to improve human resource capacity in the region to cater for tourism.

In the course of two years, the project's conclusion report suggested that both the number of young people trained on nature-based sports and female participation in the work force had increased. The accommodation's services quality had also improved. Community involvement in the project was good. Tourist movements to the region increased, with more than 500,000 people having visited the region in 2010.

Methods

Study site and the participants

This study aimed to examine the impacts of the earliest CBT projects in Turkey, namely Eastern Anatolia Tourism Development Project conducted in Coruh Valley, as perceived by accommodation business operators in the region. Therefore, our fundamental research question was how did the accommodation business operators perceive the impacts of the project. The study site was the Coruh Valley, and the study frame was the accommodation establishments operating in it. The list of these businesses was accessed on the webpage of the project (www.choruh.com.tr). A total of 46 businesses were determined, and they comprised the sample of our study.

Data collection

A semi-structured interview technique was employed for the purpose of data collection. Semi-structured interviews are important data sources for qualitative research and are generally organized around a set of predetermined open-ended questions, with other questions that arise during the dialogue between the interviewer and interviewee/s (DiCicco-Bloom & Crabtree, 2006). In the same line, the individual in-depth interview allows the interviewer to investigate personal matters in depth (Tsaur & Lin, 2014: 30). In this kind of data collection, encouraging the willingness of the participants to take part in the survey is crucial, and therefore they should be ensured about the confidentiality of the interviews.

Accommodation business managers were selected on the basis of purposive sampling. In purposive sampling, researchers use their own judgements to select the informants who are representative of the sample population (Yolal, 2016: 80). An extensive literature review was carried out to formulate the open-ended questions. They are as follows:

- How do you evaluate the tourism development in Coruh Valley, comparing before and after the project?
- Do you think you have gained the expected social and economic benefits from the project?
- Were there any negative impacts on the environment and the community during the project?
- What should be done in Coruh Valley to maintain and sustain tourism development? What are your suggestions?

Participants were informed about the study and the process was explained prior to interviews. Interviews were conducted from 15th to 30th June 2018 by phone calls. Demographic profiles of the participants and their replies to the study questions were recorded, and transcribed. A descriptive analysis was made on the data. Data collection was continued until no new information appeared which is termed as data saturation (Morse, 2004). A total of 19 participants were interviewed.

Analysis

Data were analysed descriptively. Descriptive analysis is a technique to summarize the data according to pre-determined themes. In this kind of analysis, the researcher frequently refers to direct citations from the interview transcripts to highlight the results. The main purpose is to summarize and present the results. Descriptive analysis has four stages. In the first stage, the researcher sets a frame on the basis of research questions, theoretical background or interview results. As such, the themes in which the data will be organized and presented are determined. Further in the second stage, the researcher reads the data and organizes them according to the themes. In this process, it is important to group the data in a meaningful and logical order. In the next stage, the researcher defines the organized data. This can be accompanied by direct quotations. Finally, the researcher explains the findings and relations. The researcher can also strengthen the comments and discussions by explaining the relations among themes.

Results

Demographic profiles of the participants showed that most of them were male (15) with 14 married. They were mostly high school graduates (9). Only one person had a university degree. Ten participants obtained financial funds from the project and started their own boarding houses.

The first question of the study asked participants to evaluate tourism development in Coruh Valley considering before and after the project. Participants mostly had positive attitudes towards the developments experienced during and after the project. The main benefits of CBT are the direct economic impacts on families, socioeconomic improvements and sustainable diversification of lifestyles (Manyara & Jones, 2007). They agreed that this CBT development project contributed greatly to the social and economic development in the valley. In this regard, for example one of the participants (P3) noted that:

> 'The Project has been a hope for the people in the region. In the past, there were very few boarding houses in the region, and no one travelled here. With the Project, the number of visitors to the region has increased dramatically, and the local people became familiar with tourism'.

CBT is characterized by small-scale enterprises with strong ties to other local industries, and is human centred (Tolkach & King, 2015). The participants claimed that the project had created an entrepreneurial drive among local people. This resulted in the launch of several service businesses. In this vein, participant (P6) detailed his business life:

> 'The Project encouraged us to make use of our properties which were once useless. I was the first to convert my house into a boarding house. Many people did the same. I am 70 years old, and I closed the boarding house since I am alone, but other people are making good money'.

CBT projects are related to small rural communities and nature conservation through ecotourism, the concept of which has been extended to a range of different tourism products (Zapata et al., 2011). Participants claimed that natural, historical and cultural resources of Coruh Valley were unproductive before the project. The project triggered an awareness of these resources. Accordingly, one of the participants (P5) stated:

> 'Coruh Valley is among the richest regions of the world in terms of biological diversity. People had less information about this fact before the project. The project promoted our region and informed people about its unique features. Consequently, people from all around the world visit our region'.

It was also seen that the project diversified tourism offers in the region. A bundle of tourism products flourished such as trekking, cycling, rafting, canoeing, mountaineering, camping and bear-, butterfly- and birdwatching. The emergence of these tourism products also increased the interest among tourists to visit the region. Moreover, a significant increase in the number of visitors was recorded by the project. For example one of the participants (P19) explained that: 'The Project developed the infrastructure for ecotourism. There was no accommodation in Uzundere. But now we host more than 1000 guests a year, and 100% of our guests are satisfied with their stay'.

Intangible heritage of the region was also brought to surface as a result of increasing visitation. This was found to be initiated by the training efforts during the project. One of the participants (P7) underlined this fact:

> 'The Project protected not only our natural environment but also our culture. Women were trained about gastronomy, cooking, serving and hygiene. Consequently, an inventory of forgotten tastes and recipes was rediscovered. We are happy that our food culture is protected by the project'.

In line with the Social Exchange Theory (Ap, 1992), it was found that people who were directly supported by the project have developed a positive attitude towards it. On the other hand, the ones who had no financial incentives or funds had negative attitudes. One of the participants (P8), commented:

> 'I have been operating a small family business (boarding house) since 1995, which means I was here even before the project. The project did not offer funds or support for the existing businesses to increase their capacity. I wish I could get technical or financial support'.

Unemployment has been a drastic problem for Turkey, especially in rural areas. As such, CBT development projects have the potential to increase employment opportunities in rural areas. This was verified by the participants. They stated that the project created employment opportunities for women, young people and seniors. However, some of the participants believed that only few people had the chance to be involved in the project. For example, one of the participants (P17) noted that: 'Not all businesses were supported by the project. Only a small number of local people were trained in the tourism industry. Therefore, the project could not benefit all the people'.

In CBT projects, community participation is seen as a convenient tool for educating locals about their rights, laws and political good sense (Tosun, 2000) and about the business practices. Participants who were assisted and trained in the project support its developments. In this regard, one of the female participants (P11) stated that:

> 'I was a housewife before the project. I had no idea of what tourism is. I started a boarding house with the support of the project; attended training programmes and taken to familiarization tours to established tourism regions. I learnt how to operate a boarding house. The project taught us how to catch a fish rather than receiving it. Therefore, I am happy with the project's results'.

Tourism development without proper planning can result in negative social, cultural, environmental, and economic impacts to host communities (Sheldon & Abenoja, 2001). Therefore, participants were asked about any negative impacts on the environment and the community during the project. Almost all the participants believed that the project has neither negative impacts on the environment and the community, nor on the quality of life. For example, one of the participants (P16) stated:

> 'Following the project, the Valley became popular. Previously, very few people visited this region but the number of visitors has increased dramatically after the project's inception. The visitors are responsible, they protect nature and respect local people and their lifestyles. Therefore, we experienced no problems'.

However, some of the participants were suspicious about the future of the project. They accepted that the project has no negative impacts yet on the country, historical buildings are restored, and the environment is protected. The main concern is the hydroelectric power plants built around the valley. One of the participants (P8) expressed his doubt as follows:

> 'There are several dams and hydroelectric power plants, and more are being built in Artvin. Construction of these plants harms our forests. Therefore, I have doubts about whether tourists will come if we lose our villages, highlands, lakes, rivers, flora and bio-diversity. The proposed construction of a hydroelectric power plant in Uzundere was aborted because of the protests from residents, local authorities and non-governmental organizations. But what will happen in other regions of the Valley?'

Finally, participants were asked their suggestions for Coruh Valley in order to maintain and sustain tourism development. Participants' responses to this question yielded several issues such as the supply of grants and incentives for new investments, organizing events that might attract visitors, increasing the number of qualified employees, decreasing conflicts among different institutes and developing collaboration. For example, the development of new events, a participant (P12) proposed that:

> 'The number and variety of events and festivals in the region has increased. Europe Rafting Championship was organized in Coruh Valley in 2010. Bird watching events are continuing. However, the region has plenty of potential for the organization of new events, and they may attract more visitors to the region'.

Some participants thought that the region is lacking in qualified employees. Moreover, they complained about the lack of professionalism in the operation of businesses. In this regard, one of them (P19) noted that:

> 'We don't have skilled and trained employees to work in tourism development. The number of business people who know cost control, food service, foreign language, information technologies etc. are very limited. Therefore, training and educating local people should be continued'.

An overall approach that would require the participation of related institutes and bodies is needed for the success of CBT development projects. However, the participants thought that there was no collaboration among the institutes in the region. In this vein, a participant (P18) noted that:

> 'Tourists are canoeing or fishing in Tortum Lake, but the constabulary officers get them out of the lake on the grounds that the lake is a first degree priority area for protection. Similarly, the local residents earning their living with fishing are prohibited from doing it. We agree that nature should be protected, however, is it not possible to do our job in a way compatible with nature? We expect collaboration and coordination of the central government and local authorities'.

Conclusion

Coruh Valley has a significant tourist potential with its natural, cultural, historical resources and recreational opportunities. This region was developed as a tourist destination by a CBT development project. This chapter examined the project by investigating business owners' perceptions and attitudes towards tourism development in the region since the inception of the project.

The findings of the study revealed that the project contributed significantly to tourism development in the region. Tourism was introduced in the region, new businesses were opened, and residents were trained. Specifically, training the women, getting them involved in economic activities and allowing them to earn their living, were among the most important outcomes of the project. Further, new employment opportunities were created and young people were encouraged to participate in tourism development. When CBT projects are evaluated, it is seen that community resources such as cooperatives, unions and educational resources are operated voluntarily by the community (MacDonald & Jolliffe, 2003; Mbaiwa, 2003; Johnson, 2010). Accordingly, the project in Coruh Valley aimed to use the community resources to attract tourists into the region. The findings also showed that the participation of the local people was encouraging, and the benefits were divided among the community. In line with literature, people who had been engaged in tourism and obtained personal benefits were more supportive of the project. Therefore, it is imperative to expand the involvement of local people and create ways to increase economic and social benefits of the CBT development projects. Findings further revealed that the project had increased the number of hospitality businesses in the region. Accordingly, previously abandoned buildings were restored and operated as boarding houses or accommodation units. Services quality has also improved.

Overall the study revealed that the project has a significant influence on the development of tourism in the region. Expectedly, it also contributed to a better quality of life among residents in the valley. However, since the project lasted for a limited period of four years, further efforts are needed to sustain what has been achieved. Similarly, as suggested by the study findings, the influence of the project was limited and just a limited number of people benefitted. As noted by Liu (2003) and McKercher and Prideaux (2014), CBT is either displaying slow progress or failed to demonstrate tangible contributions. Future projects should aim to expand the benefits of tourism development and increase the number of people involved in tourism (Karacaoğlu et al., 2016). This is specifically needed to increase the support from local people. Poverty alleviation will increase the quality of life which eventually results in better support from local residents for tourism development.

Although the CBT project has brought about economic benefits to sections of the local community, its sustainability beyond the 4-year time frame will be challenged by conflicting uses notably the probable construction of dams and hydroelectric plants. Having enjoyed the initial success of tourism, the resilience of the local community will be tested in upscaling the CBT project in line with the principles of rural responsible tourism. Credit should be given to the project proponents for rehabilitating the historical buildings that has improved placemaking within the destination. However, the local stakeholders' overall understanding of the concept needs to be harnessed to ensure 'buy in', which is reflected in the contestations over the use of Tortum Lake.

In essence, the CBT pilot project at Coruh Valley can be said to have achieved its main aim of rural economic revitalization in line with Targets 1 and 8 of the United Nations Sustainable Development Goals (SDG 1, 'No Poverty' and SDG 8, 'Decent Work and Economic Growth' respectively). What remains to be seen is where the CBT project could trigger intangible benefits such as sense of pride, community cohesion, and self-esteem, etc. Essentially these values are necessary for the local community to strengthen their resilience when the 'handholding' by the UNDP project ends – herein lies the start of the more challenging part of sustaining CBT.

By land mass, about 95% of Turkey is located in Asia and the Coruh Valley is situated in Anatolia or what is previously called Asia Minor. As highlighted earlier, the region is blessed with an abundance of natural and cultural resources but the people are not aware of these assets. The cultural landscape of the Coruh Valley has also been shaped by centuries of human intervention but in a way that blends and respects the natural environment. The UNDP project has focused primarily on soft adventure tourism that is suited to the terrain and physical environment of the region. Once the local community has reached a capacity that can systematically handle responsible rural tourism, it should present a more holistic tourism offering and experience based on the intrinsic harmony between human and nature as the rural imagery.

References

Ap, J. (1992) Residents' perceptions on tourism impacts. *Annals of Tourism Research* 19 (4), 665–690.

Blackstock, K. (2005) Critical look at community based tourism. *Community Development Journal* 40 (1), 39–49.

Coruh (2018) Main page. See http://www.choruh.com.tr (accessed 3 April 2018).

DATUR Final Presentation (2011) See http://www.choruh.com/tr/datur/reports (accessed 12 April 2018).

DiCicco-Bloom, B. and Crabtree, B.F. (2006) The qualitative research interview. *Medical Education* 40 (4), 314–321.

Goodwin, H. and Santilli, R. (2009) Community-based tourism: A success. *ICRT Occasional Paper* 11 (1), 37.

Hiwasaki, L. (2006) Community-based tourism: A pathway to sustainability for Japan's protected areas. *Society and Natural Resources* 19 (8), 675–692.

Johnson, P.A. (2010) Realizing rural community based tourism development: Prospects for social-economy enterprises. *Journal of Rural and Community Development* 5 (1), 150–162.

Karaçaoğlu, S., Yolal, M. and Birdir, K. (2016) Toplum temelli turizm projelerinde katılım ve paylaşım: Misi Köyü örneği. *Çağ Üniversitesi Sosyal Bilimler Dergisi* 13 (2), 103–124.

Kibicho, W. (2008) Community-based tourism: A factor-cluster segmentation approach. *Journal of Sustainable Tourism* 16 (2), 211–231.

Kithir, P. (2012) *Gelecek turizmde Çoruh Vadisi deneyimi*. See http://kurumsal.data.atilim.edu.tr/pdfs/121212.pdf (accessed 16 May 2015).

Lapeyre, R. (2010) Community-based tourism as a sustainable solution to maximise impacts locally? The Tsiseb Conservancy case, Namibia. *Development Southern Africa* 27 (5), 757–772.

Liu, Z. (2003) Sustainable tourism development: A critique. *Journal of Sustainable Tourism* 11 (6), 459–475.

López-Guzmán, T., Sánchez-Cañizares, S. and Pavón, V. (2011) Community-based tourism in developing countries: A case study. *Tourismos: An International Multidisciplinary Journal of Tourism* 6 (1), 69–84.

MacDonald, R. and Jolliffe, L. (2003) Cultural rural tourism: Evidence from Canada. *Annals of Tourism Research* 30 (2), 307–322.

Manyara, G. and Jones, E. (2007) Community-based tourism enterprises development in Kenya: An exploration of their potential as avenues of poverty reduction. *Journal of Sustainable Tourism* 15 (6), 628–644.

Mbaiwa, J.E. (2003) The socio-economic and environmental impacts of tourism development on the Okavango Delta, North-Western Botswana. *Journal of Arid Environments* 54 (2), 447–467.

McKercher, B. and Prideaux, B. (2014) Academic myths of tourism. *Annals of Tourism Research* 46, 16–28.

Morse, J.M. (2004) Theoretical saturation. In M.S. Lewis-Beck, A. Bryman and T.F. Liao (eds) *The Sage Encyclopedia of Social Science Research Methods* (p. 1123). Thousand Oaks, CA: Sage. See http://sk.sagepub.com/reference/download/socialscience/n1011.pdf (accessed 1 December 2017).

Okazaki, E. (2008) A community-based tourism model: Its conception and use. *Journal of Sustainable Tourism* 16 (5), 511–529.

Salazar, N.B. (2012) Community-based cultural tourism: Issues, threats and opportunities. *Journal of Sustainable Tourism* 20 (1), 9–22

Sebele, L.S. (2010) Community-based tourism ventures, benefits and challenges: Khama Rhino sanctuary trust, central district, Botswana. *Tourism Management* 31 (1), 136–146.

Sheldon, P.J. and Abenoja, T. (2001) Resident attitudes in a mature destination: The case of Waikiki. *Tourism Management* 22 (5), 435–443.

Tolkach, D. and King, B. (2015) Strengthening community-based tourism in a new resource-based island nation: Why and how? *Tourism Management* 48, 386–398.

Tosun, C. (2000) Limits to community participation in the tourism development process in developing countries. *Tourism Management* 21 (6), 613–633.

Tsaur, S.-H. and Lin, W.R. (2014) Hassles of tour leaders. *Tourism Management* 45, 28–38.

United Nations Development Programme (UNDP) (2018) Welcome to Uzundere, the Secret Gem of Çoruh Valley. See http://www.tr.undp.org/content/turkey/en/home/ourwork/povertyreduction/success/stories/welcome-to-uzundere—the-secret-gem-of-coruh-valley.html (accessed 10 May 2018).

Yolal, M. (2016) *Turizm araştırmalarında örnekleme-bibliyometrik bir araştırma* (Birinci Baskı). Ankara: Detay Yayıncılık.

Zapata, M.J., Hall, C.M., Lindo, P. and Vanderschaeghe, M. (2011) Can community-based tourism contribute to development and poverty alleviation? Lessons from Nicaragua. *Current Issues in Tourism* 14 (8), 725–749.

9 Community Characteristics, Social Cohesion and the Success of Community-Based Tourism: Case Studies of Vietnam

Tramy Ngo and Nguyen Thi Huyen

Introduction

Community-based tourism (CBT) has been widely acknowledged as one of the alternative forms of tourism that orient towards sustainable development. CBT offers the potential to contribute to the objectives of poverty alleviation, community empowerment, preservation of natural and cultural resources and diversification and improvement of the tourism experience (Hiwasaki, 2006; Manyara & Jones, 2007; Salazar, 2012). Nevertheless, such benefits of CBT initiatives are hardly achievable in practice. Indeed, only a few CBT projects maintain their viability in the long term and a majority of CBT projects collapse after a funding period (Goodwin & Santilli, 2009; Mitchell & Muckosy, 2008; Rocharungsat, 2008). In Vietnam, CBT development dates back to the years of the 2000s, when the central government identified tourism as a key sector for national employment and economic growth. Under the facilitation of the government, international and local non-governmental organizations (NGOs) supported the development of CBT initiatives, particularly in the rural and mountainous regions of the country. However, like many other parts of less economically developed countries, the sustainability of CBT initiatives in Vietnam is questionable.

The literature of CBT has documented numerous attempts undertaken to investigate factors shaping the long-term success of CBT as well as reasons causing its failure. Among others, community-related characteristics play a crucial role in initiating a CBT project and fostering its sustainability. Different community characteristics result in different incentives affecting community involvement in tourism activities and their resource management. According to Barrow and Murphree (2001), four community-related criteria that influence the success of conservation objectives are cohesion, legitimacy, delineation and resilience. Within the scope of this chapter, we focus on the importance of community cohesion in regulating the success of CBT projects. In particular, through the ethnographic and interviewing techniques and two case studies of My Son CBT and Triem Tay CBT (Vietnam), we argue that historical homogeneity of community, kinship linkages and cultural sharing help to improve social cohesion among community members. Community cohesion owing to their historical homogeneity, kinship linkages and indigenized cultural sharing may stimulate the positive responses of community members towards the paradox of individualism and collectivism associated with community interventions in a CBT project, thereby influencing its sustainability.

Community Characteristics and Community Cohesion in CBT Development

Community cohesion and sustainable CBT are intertwined. Cohesion, according to Barrow and Murphree (2001: 26), refers to 'a sense of common identity and interests which serves to bring people together for collaborative action, and leads to collectively differentiate themselves from others'. Community cohesion regulates the willingness to set and strive for common goals within a community. Community cohesion is among community development goals that CBT endeavours target. The evaluation of CBT sustainability cannot be fulfilled without addressing the aspect of community cohesion. Indeed, CBT is expected to benefit community individuals and simultaneously encourage community cohesion and harmony (Foucat, 2002). Likewise, Taylor (2016) suggested the inclusion of social cohesion together with the conventional triple bottom line metric, which accounts for ecological health, financial sustainability and local social capital, in evaluating CBT success. Concomitantly, community cohesion is a source of social capital, of which improvement or lack of this resource would affect the long-term success of a CBT project. A lack of community cohesion may hinder the accumulation of trust and reciprocity, which are the components of cognitive social capital, among community residents (Jones, 2005). Farrelly (2011) argues for the integration of social empowerment, a situation in which tourism initiatives promote social cohesion and integrity, in the process of indigenous tourism development.

Among other factors, community characteristics shape its social cohesion (Barrow & Murphree, 2001). The community characteristics are configured through the community values perceived by its members. Such perceived community values shape community pride, reinforce community resilience and regulate community identity. In particular, a community with consistent characters, resounded community identity, and high community resilience is arguably more proactive in tourism involvement, and less vulnerable to tourism hazards. A study by Taylor (2016) indicated that elite domination and kin group control, which are the sociocultural aspects within a community, stimulated different levels of participation in CBT, thereby exacerbating existing social tensions and spoiling community cohesion among community members in a CBT project in Mexico. Likewise, Farrelly (2011), through sociocultural and political viewpoints, pointed out the impact of kin groups, village spokesmen and clan systems on democratizing decision-making processes among community members while participating in ecotourism projects in Fiji. Thus, it is crucial that the members of a community acknowledge community values and be aware of their embedment in the community to promote community cohesion, and contributing to the long-term success of CBT development.

It is worthwhile canvassing the linkages between community characteristics and community cohesion in CBT sustainability. Such research may contribute to providing an evidence-based reference for CBT planning and development, customize CBT initiatives for particular communities, thereby better promoting CBT sustainability. In terms of knowledge contributions, an attempt on investigating the relationships between community characteristics and community cohesion can elucidate the variable of social cohesion, thereby contributing to a more holistic approach in CBT evaluation (Taylor, 2016; Whitford & Ruhanen, 2014). Accordingly, attributes included in community characteristics are identified to provide an analytical framework for the study. Under a resource management perspective and in an orientation toward community cohesion, the community characteristics encompass historical factors (population and settlement history; conflict history), social factors (ethnicity and language; family structure; caste and other social divisions; gender relations), economic factors (differences or similarities in livelihood strategies; the degree of economic stratification in the community) and cultural factors (religion, cultural beliefs) (Thomson & Freudenberger, 1997). Given the overarching of the objective of resource management on community involvement in tourism, we adopted this terminology in our chapter. Indeed, through the analysis of two case studies using ethnographic methods, we argue that historical homogeneity, kinship linkages and cultural sharing of a community can diversify the level of community cohesion in CBT development. Particularly, this chapter is aimed at interrogating the social and historical aspects of community cohesion, which is a relatively less explored knowledge-scape within the CBT literature.

The Study Context: My Son and Triem Tay CBT Villages (Vietnam)

My Son CBT village

My Son village is located in DuyPhu Commune, DuyXuyen District, Quang Nam Province, Vietnam. The village is at the edge of the My Son

World Heritage archaeological site, renowned as the holy land of Cham civilization from the 6th to 14th centuries. In 1999, the My Son Sanctuary Heritage site was recognized by United Nations Educational, Scientific and Cultural Organization (UNESCO) as the Cultural World Heritage Site. Being part of the governmental policy of establishing buffer zones surrounding World Heritage sites, My Son village has been resettled. The village, whose habitats originate from different regions of the province, has more than 50 households. They mostly earn their living by farming and cattle herding. Regardless of thousands of visitors visiting My Son Sanctuary Site annually, local communities surrounding this attraction, including My Son villagers, have not benefited from tourism. In addition, due to the preservation policies of World Heritage Sites, forests surrounding the My Son Sanctuary Site were preserved against local access for cattle herding. These issues motivated the initiation of CBT development in My Son CBT village in 2011.

The initiation of My Son CBT village is associated with the project of International Labour Organisation (ILO) 'Strengthening Inland Tourism in Quang Nam'. The project aimed to particularize the model of poverty reduction through tourism development and promote economic growth at the grassroots level using local resources. With technical and financial assistance from ILO, which lasted from 2011 to 2016, tourism activities in My Son CBT village were initiated and developed. In particular, the My Son Community Tourism Cooperative was established in March 2013. The Cooperative has 23 members, being managed by a management board. Concurrently, various tourism services such as homestays, food services, agricultural experiences, hiking and boating were designed and commercialized. Various external stakeholders were engaged to support the Cooperative operation and the viability of My Son CBT village. In particular, the management board of My Son Sanctuary Heritage Site assisted the Cooperative in connecting the village to potential tourists. TraKieu tour operator was invited, through public–private contractual partnerships, to facilitate market accesses for CBT activities. The local government committed itself in supporting the village in road building, electricity and water supply. ILO, adding to the role of a facilitator connecting the village to other abovementioned stakeholders, also provided the village and the Cooperative with technical support, including capacity-building training and business-skills training.

Triem Tay CBT village

Triem Tay, a beautiful riverside village, is located in Dien Phuong commune, Dien Ban District, Quang Nam Province, Vietnam. The village is about a 10-minute boat ride from Hoi An – a tourist hub and among the top appealing tourist attractions in Vietnam. Different from My Son village, which is re-settled through re-dwelling programmes, Triem Tay has built its traditions, history and culture over hundreds of years of development. Nevertheless, the village has been almost excluded from tourism evolutions in the region. In particular, owing to tourism growth in Hoi An, and in Triem Tay's neighbouring villages of Tra Que, Thanh Ha which have been widely promoted to attract visitors. Numerous reasons are identified to explain this tourism negligence. Triem Tay, due to its closeness to Thu Bon River, has its landscape seriously threatened by land erosion. Also, the village suffers from severe floods annually. Additionally, the traditional livelihood of the residents, which is mat weaving, has become obsolete due to the prevalence of industrialized products. Only one or two families in the village continue their weaving work. Thus, the lack of land for farming and grazing in the village is exacerbated by the stagnation of traditional crafts. Consequently, there is a severe shortage of livelihoods for the village residents. As a result, the issue of emigration has occurred in Triem Tay over the years, in which young citizens have left the village and immigrated to nearby cities for jobs.

In 2014, Triem Tay was selected to initiate tourism interventions under national-targeted programmes 'New Rural Development' and 'One Village-One Product'. The interventions were initially proposed by the local government and were facilitated through collaboration between UNESCO and ILO. This cooperative effort aimed to employ tourism potential to create jobs, reduce poverty and develop satellite destinations around Hoi An World Heritage Site. Adopting a participatory approach, Triem Tay residents and local authorities at the communal and district levels worked together to develop action-based plans for tourism development. In particular, the government collaborated in strategies specifically aimed at preventing land erosion, supplying water and electricity and improving road conditions in the village. In 2015, with the support of the Provincial Alliance of Cooperatives, the Triem Tay Agri-Ecotourism Cooperative was established, consisting of 33 households. Currently, tourism activities offered in this village and under the management of the Cooperative include daily tours of village sightseeing, boating, weaving and food offerings.

The selection of these two case studies is justified. These two villages share three characteristics in common. First, with the help of ILO, the two villages have recently developed CBT initiatives. Second, the villagers are Kinh people, who represent the majority ethnic group in Vietnam. Third, the villages are close to World Heritage Sites, which help to secure potential markets for CBT activities. However, these two villages differ in their ethnic homogeneity and historical prestige. In particular, the village of Triem Tay has a history of more than 400 years, while My Son is a resettled village with more than a 20-year history. The similarity and differences between these two villages are highlighted to indicate how community characteristics of ethnic homogeneity and historical prestige could affect the degree to which community cohesion is altered.

Table 9.1 Informants as per the case study

Village	CBT village members	CBT Cooperative Management Board	Tour operators	Tourism government
My Son	05	01	01	01
Triem Tay	10	01	–	01

Methodology

The ethnography method framed the methodology of this chapter. This method has been widely employed to investigate community-related issues in indigenous tourism research (Koot, 2016; Ruiz-Ballesteros & Hernández-Ramírez, 2010; Taylor, 2014). In particular, the second author of this chapter was thoroughly engaged in the CBT development in the two villages in the role of a project coordinator. The results presented in this chapter are mainly based on informal interviews and observations that she conducted in the field trips from 2011 to 2016. Interview informants included the villagers, head of the villages, the director of tourism cooperatives, local governmental officers working in charge of cultural issues of the district, the tour operators working with the tourism cooperatives. In total, 20 informants were interviewed within the abovementioned timeframe. The informants were asked about which historical–cultural factors, as per their perception, contribute to regulate their embedment to the community. They were also asked how those factors affect the tourism development in the community. Table 9.1 shows the informants per case studies.

In addition, the technique of participant observation was also employed, owing to the engagement of the researchers in the two case studies in the period from 2011 to 2016. These data were collected in the form of diaries, insights and notes through the period. The data were then analysed manually using content analysis.

Findings

Community characteristics perceived by community members

Two different perceptions about community values were recorded through interviewing residents of the two villages. In particular, one stream perceived community values as indistinctive, vague and with low connections (reflected in the case of My Son CBT village) whereas the other stream valued their community characteristics as consistent, unique and a privilege to be member of the community (demonstrated in the case of Triem Tay CBT village).

According to My Son villagers, their village is identified through cultural values, which are mainly sourced from typical Vietnamese culture and the Cham culture of My Son Sanctuary Heritage site (Cham culture derives its spiritual origins from the Hinduism of India and can be traced to the people of the Champa Kingdom in the 4th to 13th centuries (UNESCO, 1999)). Inhabited by the Kinh people, My Son CBT village possesses the cultural attributes of a typical Vietnamese countryside village. These cultural resources include the friendliness of hosted communities and a fresh and healthy cuisine. A villager said, 'Friendly smiles help us to welcome visitors' (Ms V, a homestay operator in the CBT cooperative). However, this village's members could not see their village as being different from other nearby communities. An informant stated, '[Our village] is similar to the other villages in terms of our customs and practices, we do not see ourselves as being special because we are all the Kinh people' (Mrs A, a primary school teacher and a member of CBT Cooperative). Additionally, the Cham culture of My Son Sanctuary Heritage site defines the cultural authenticity of the site and its surrounding destinations, including My Son CBT village. Unfortunately, none of this culture's trails is represented in their liveable forms within the village because the residents dwelling in the village are non-locals through re-settlement programmes. Also due to the short history of village establishment, My Son village does not have any Village Saints or Protectors who, as per the belief of Vietnamese culture, can protect villagers from natural hazards and inspire the village's prosperity and peace. A lack of those spiritual facilitators resulted in discursive cultural sharing among the residents of My Son CBT village. No annual festivals or communal workshop places have been specified for the village.

Community characteristics illustrated in the Triem Tay village represent a different viewpoint. The community identity is well defined through a long-standing history, kinships among community members, and typical cultural values. For instance, the village has been renowned for its mat weaving since the 18th century. A villager commented, 'We are proud of our glorious history. Our village used to be renowned for its mat products. Our village was also a stop-over place for the Kings of Nguyen Dynasty (on their travel to the South [Hai Van pass Southward]) in the 18th and 19th centuries' (statement from Mr Y, one of the oldest men in the village). Additionally, around 150 households reside in the village, whose kinship networks can be traced back to over hundreds of years. In addition to Vietnamese cultural features, Triem Tay also possesses unique cultural values, for instance, folk songs. 'We are also proud of our culture. Our folk songs make us distinct from other nearby villages' (Mrs T, one of the village's folksong performers). Also, in Triem Tay, a chapel has been built for worship purposes. Annually, the village's oldest men organize events to show their respectfulness toward the village's Protectors at this chapel.

The embedment of community members to the community

History and cultural values mould the embedment of community members in their community. The difference in history and cultural values differentiates the commitment of the citizens of My Son and Triem Tay toward their village and regulates different levels of social cohesion among them. Low cohesion and poor connections reflect the embedment of My Son villagers to the community, whereas community pride and firm embedment represented the linkages between Triem Tay villagers and their community.

At My Son CBT village, a lack of blood relationships among community members and a short historical conduit of the village arguably resulted in the low embedment of members to the community. Due to resettlement programmes, the villagers, who originated from different parts of the country, did not have any common historical or cultural sharing. A sense of community connectedness and a perception of community values were vague among villagers. In particular, community members at this village, when asked to promote their village to tourists, were confused because they recognized that they own no indigenized cultural values rather than Vietnamese culture, which speaks for Vietnam in general, and Cham culture, which is not their cultural representative. Also, they were unaware of any historical stories linked to their village.

The connectedness to land, history, tradition and kinship linked Triem Tay citizens and their village. In Triem Tay, villagers found themselves connected to the community through family bonds to the land. A villager mentioned, 'My family have resided here over many generations, including that of my grandfather, my father and me, along which my love to this village is nurtured' (Mr V, 80-year-old villager). The embedment to the community was also reflected through blood relationships among community members. 'We know each other very well, and many of us share family trees, having the same ancestors' (Mr V).

Community pride was well presented among residents in Triem Tay village. The community pride comes from the village's unique culture and long-standing history with more than 400 years of growth. The tourist flyer about Triem Tay village, which is drafted by Triem Tay CBT Cooperative, reflects the pride among community members:

> The Triem Tay village was established at the beginning of the century XVII on the banks of Thu Bon River. The beauty of Triem Tay is unique, characterized by moorings, ferry boats on the river, rustling bamboo groves, old and rusty houses with baked tiles, and a variety of ancient temples. At Triem Tay, the locals enjoy their lives happily regardless of daily difficulties traced on their faces. Thus, visitors are induced by peaceful reflexivity along with their first steps to the village. Furthermore, travelers can discover hidden gems shining from the landscape, the people and the history of this riverside village. These appeals do not simply offer picturesque sights but also convey the unique identity of the village through its people and culture.

Owing to such pride, tourism development in the village has been enthusiastically supported and initiated by community members. In particular, the issue of emigration has been relieved owing to tourism potentials activated in the village. Mrs H, a governmental officer said, 'a number of local people left the village two years ago because of difficulties in business and jobs. Now they are planning to return home and contribute their fund to the Cooperative as well as re-settle at the village'. Mr V, a villager, had the same observation:

> 'Because of natural disasters that the village has to suffer every year such as floods and consequently land erosion, many people had to move to Da Nang city (30 km from the village) to seek jobs. They left their house unmaintained. Since the village is being revitalized through tourism, some of them returned to the village, modified their house into a homestay, a guesthouse or a weekend-get-away place. I believe that in the heart of each villager, nobody wants to leave the village'.

Community members at Triem Tay village perceived their community values as a top priority and the most prestigious social asset to engage in tourism. All the interviewees agreed that Triem Tay's tourism offerings should be mainly based on the village-owned values (i.e. folk songs, authentic cuisine, villagers' storytelling, historical relics and sites, the harmony of daily lives). The cultural appreciation was highlighted in villagers' conversations about tourism development alternatives at the village. According to Mrs H (governmental officer):

> 'When we (the local government) asked villagers to materialize their cultural values to tourism products, they proudly replied to us about the stories related to the village, about the village history and traditions. What else should a CBT traveller expect about a CBT product rather than those stories?'

CBT evaluation correlated to different levels of social cohesion

Manifest indicators of tourism growth, such as tourism revenue and tourist flows, were predominantly adopted to evaluate CBT development at a destination. At the time of drafting this chapter (i.e. September 2017), My Son CBT village had been in operation for over six years while Triem Tay village had been in operation for three years. Within one year of operation (2015–2016), the Triem Tay CBT Cooperative yielded a total income of USD50,000 from tourism activities. At My Son CBT village, in six months since its official operations (i.e. June–December, 2013), the Cooperative served 1000 lunches while the other services managed by the Cooperative such as boating and hiking were relatively less appealing. Currently, through the researchers' observations, and following the CBT lifecycle proposed by Zapata *et al.* (2011), CBT development in My Son village could be evaluated as at a stagnation stage. Contrastingly, CBT activities in Triem Tay are positioned at the stage of accelerated growth.

Conflicts associated with CBT development and conflict resolutions arguably indicate the level of community cohesion, which is either decreased or improved through involvement in tourism. In the case of My Son CBT village, two years from its inception, the paradox of individualism and collectivism occurred, and conflicts emerged, reflected in tensions among different service groups and between service groups and the Cooperative management board. Due to a discrepancy among different service groups in terms of income from tourism, community members came to regard as unfair the rotation mechanism applied to run the Cooperative. Furthermore, the perceived credibility of the Cooperative's management board was very low among the community. In particular, the Cooperative members were dissatisfied with the head of the Cooperative regarding his vague clarification between individual benefits and collective benefits in the management activities of the Cooperative. Cooperative members criticized this individual for his bias in regulating rotations and for using the Cooperative privilege to benefit his family-owned business. In general, Mr T, a tour operator promoting CBT offerings of My Son CBT village, commented:

> 'The mechanism of rotation for a fair delivery of tourism benefits does not work well in the case of My Son CBT village. In my opinion, the dominance of individualism in collective activities and the non-connectedness among villagers arguably are the main reasons causing such discrepancies and tensions among community members'.

Conflict resolutions were evaluated as ineffective, which consequently exacerbated the level of social cohesion among community members. Indeed, conflicts associated with CBT management in the village were raised in an official meeting led by a local authority. As this local authority official was not involved in the CBT Cooperative Management Board, he was regarded as an *outsider*. The perception of an *outsider* attempting to resolve a conflict-solving meeting hindered community members from speaking out their conflicts, thereby worsening tensions. As a result of these tensions and ineffective conflict resolutions, polarities in the village's CBT management system occurred. Currently, there are two other self-established cooperatives in parallel to the original CBT Cooperative participating in CBT involvement at the village. Concomitantly, CBT development in the village has significantly stagnated, as illustrated previously.

In contrast, the community cohesion has been well reinforced through tourism participation in Triem Tay village. The CBT Cooperative, since its inauguration, remains as the only community representative entity in the village. There has been a continuous increase in villagers voluntarily registering to be Cooperative members. Additionally, those villagers, who do not join the Cooperative, still support tourism activities in the village. A villager stated: 'All villagers know how to do tourism businesses. Although many villagers could not be Cooperative members because of their commitment to jobs, other than tourism, they still support the development of the Cooperative' (Mr Y, the manager of the CBT Cooperative). Additionally, any conflicts that arose were resolved through night meetings among community members rather than being chaired by an *outsider*. The head of the Triem Tay village affirmed, 'we try to encourage villagers resolve conflicts themselves rather than having a meeting with any third party'.

Conclusion

This chapter is aimed at interrogating sustainable development of CBT through a sociocultural lens. Through this chapter, we argue that historical and sociocultural factors contribute to shaping community cohesion and thereby CBT development. In particular, community cohesion is reinforced in those communities, whose members share historical homogeneity, kinship linkages and community-owned cultural attributes. On this baseline, community cohesion is utilized as a sort of social capital while participating in tourism activities and concomitantly the community tension is minimized during CBT implementation. The case of Triem Tay CBT village demonstrates this argument. In contrast, in those communities that lack historical homogeneity, kinship linkages and indigenized cultural attributes, the level of community cohesion is relatively low, and this sociocultural factor is more vulnerable to tourism development. The case of My Son CBT village in this chapter is an example.

Ngo *et al.* (2018), in a study of investigating factors contributing to a successful marketing collaboration among CBT stakeholders, found that the dilemma of individualism versus collectivism in a community leader's engagement in CBT activities arouses skepticism about leaders' legitimacy, thereby affecting the viability of CBT initiatives. This chapter elaborates the argument of Ngo *et al.* (2018) through experimental research. Indeed, findings from this chapter revealed that an unclear clarification of community leaders regarding individual benefits in collective efforts causes tensions among community members, degrades community cohesion and adversely impacts the long-term viability of a CBT project. In a different aspect, kinship is argued as a sociocultural factor affecting the participation in CBT, the delivery of tourism benefits and consequently the evaluation of CBT success (Taylor, 2016). This chapter further elaborates the significance of kinship in CBT development, through its influence on enhancing or degrading community cohesion. Likewise, indigenized cultural sharing within a community is regarded as a significant attribute to be considered in CBT planning and development. Farrelly (2011), in his study, discusses the significance of *vanua*, a Fijian cultural concept and way of life, in community governance system and social bonds among community members. As per Farrelly (2001), this sort of linkage regulated the sustainable development of a community-based ecotourism project in

Fiji. In this chapter, the shared values of long-standing history and indigenous culture are argued to build up the community identity, foster community pride and enhance community cohesion, altogether positively affecting the success of a CBT project. Central to community cohesion is harmonious living with the physical environment. In what could be considered as a typical Asian rural setting, living under the threat of natural disasters such as flooding requires a harmonious relationship that is based on respect for the forces of nature. It could be surmised that Triem Tay CBT village has a soul shaped by its historical homogeneity, kinship linkages and indigenized cultural attributes, which in turn, has been instrumental in creating an authentic tourist experience. In addition to the above-mentioned knowledge contributions, findings from this chapter arguably provide an evidence-based reference for the planning and development of CBT in Vietnam and other similar contexts in order to minimize the threat of CBT failure. Our conclusion is that endeavours for CBT sustainability need to be socioculture-related and community-oriented.

References

Barrow, E. and Murphree, M. (2001) Community conservation: From concept to practice. In *African Wildlife and Livelihoods: The Promise and Performance of Community Conservation* (pp. 24–37). Oxford: James Currey Ltd.

Farrelly, T.A. (2011) Indigenous and democratic decision-making: Issues from community-based ecotourism in the Boumā National Heritage Park, Fiji. *Journal of Sustainable Tourism* 19 (7), 817–835.

Foucat, V.A. (2002) Community-based ecotourism management moving towards sustainability, in Ventanilla, Oaxaca, Mexico. *Ocean and Coastal Management* 45 (8), 511–529.

Goodwin, H. and Santilli, R. (2009) Community-based tourism: A success. *ICRT Occasional Paper* 11 (1), 37.

Hiwasaki, L. (2006) Community-based tourism: A pathway to sustainability for Japan's protected areas. *Society and Natural Resources* 19 (8), 675–692.

Jones, S. (2005) Community-based ecotourism: The significance of social capital. *Annals of Tourism Research* 32 (2), 303–324.

Koot, S.P. (2016) Contradictions of capitalism in the South African Kalahari: Indigenous Bushmen, their brand and baasskap in tourism. *Journal of Sustainable Tourism* 24 (8–9), 1211–1226.

Manyara, G. and Jones, E. (2007) Community-based tourism enterprises development in Kenya: An exploration of their potential as avenues of poverty reduction. *Journal of Sustainable Tourism* 15 (6), 628–644.

Mitchell, J. and Muckosy, P. (2008) *A Misguided Quest: Community-based Tourism in Latin America*. See https://assets.publishing.service.gov.uk/media/57a08bd2e5274a27b2000d9d/tourism-OpPaper.pdf (accessed 30 June 2016).

Ngo, T., Hales, R. and Lohmann, G. (2018) Collaborative marketing for the sustainable development of community-based tourism enterprises: A reconciliation of diverse perspectives. *Current Issues in Tourism* 1–18.

Rocharungsat, P. (2008) Community-based tourism in Asia. In G. Moscardo (ed.) *Building Community Capacity for Tourism Development*. Wallingford: CABI Publishing.

Ruiz-Ballesteros, E. and Hernández-Ramírez, M. (2010) Tourism that empowers? Commodification and appropriation in Ecuador's Turismo Comunitario. *Critique of Anthropology* 30 (2), 201–229.

Salazar, N.B. (2012) Community-based cultural tourism: Issues, threats and opportunities. *Journal of Sustainable Tourism* 20 (1), 9–22.

Taylor, S.R. (2014) Maya cosmopolitans: Engaging tactics and strategies in the performance of tourism. *Identities* 21 (2), 219–232.

Taylor, S.R. (2016) Issues in measuring success in community-based Indigenous tourism: Elites, kin groups, social capital, gender dynamics and income flows. *Journal of Sustainable Tourism* 25 (3), 433–449.

Thomson, J.T. and Freudenberger, K.S. (1997) *Crafting Institutional Arrangements for Community Forestry*. Rome: FAO – Food and Agriculture Organization of the United Nations.

United Nations Educational, Scientific and Cultural Organization (UNESCO) (1999) My Son Sanctuary. See https://whc.unesco.org/en/list/949 (accessed 9 August 2018).

Whitford, M. and Ruhanen, L. (2014) Indigenous tourism businesses: An exploratory study of business owners' perceptions of drivers and inhibitors. *Tourism Recreation Research* 39 (2), 149–168.

Zapata, M.J., Hall, C.M., Lindo, P. and Vanderschaeghe, M. (2011) Can community-based tourism contribute to development and poverty alleviation? Lessons from Nicaragua. *Current Issues in Tourism* 14 (8), 725–749.

10 Conversations with the Local Champions of Miso Walai Homestay: Responsible Tourism in Practice

Amran Hamzah

Introduction

This chapter will present the findings from a longitudinal research carried out at Miso Walai Homestay from 2005 to 2017. Miso Walai Homestay is one of the most successful community-based tourism (CBT) projects in Malaysia and has contributed to uplift the standard of living of the local community and eradicate poverty. More importantly, Miso Walai Homestay is a success story in responsible rural tourism that is based on synergy between ecotourism and conservation that has the potential of revitalizing the harmonious relationship between humans and nature.

Community-Based Tourism in the Context of Responsible Tourism

Responsible tourism is 'about making better places for people to live in and better places for people to visit', that 'aims at minimizing negative economic, environmental and social impacts' (Goodwin n.d). Mihalic (2016) added that responsible tourism directly addresses tourism behaviour and involves dialogue, creating solutions and acting to make tourism more sustainable. Responsible tourism has also been defined as a behavioural trait that is based on the principles of respect for others and their environment (Leslie, 2012); and herein, it is linked to sustainability initiatives (Chettiparamb & Kokkranikal, 2012).

CBT is a subset of rural tourism which is increasingly becoming popular in developing countries as a vehicle for revitalizing as well as stimulating the rural economy (Hjalager, 1996; MacDonald & Jolliffe, 2003; Sharpley, 2002; Tooman, 1997). Central to the characteristics of CBT are local involvement, control and benefits (Mitchell & Ashley, 2010; Scheyvens, 2002; Tosun, 2000). As CBT is usually small scale and organic in terms of its development, it resonates well with the ideals of responsible tourism at the destination level – especially the assertion that responsible tourism is to use tourism rather than to be used by it (Goodwin, 2012) This chapter will present and discuss the case of Miso Walai Homestay as a form of responsible rural tourism, which has been saved from being a poverty-induced environmental abyss to become one of the most successful CBT programmes in Malaysia.

Background

Miso Walai Homestay is a CBT project located in Mukim (subdistrict) Batu Puteh, in the Kinabatangan district of Sabah, Malaysia (Figure 10.1). The CBT project is situated along the Kinabatangan River in one of Malaysia's biodiversity hotspots. The Kinabatangan flood plains contain more than 250 species of birds including eight of the ten species of hornbills found in the country. It is also home to over 90 species of mammals

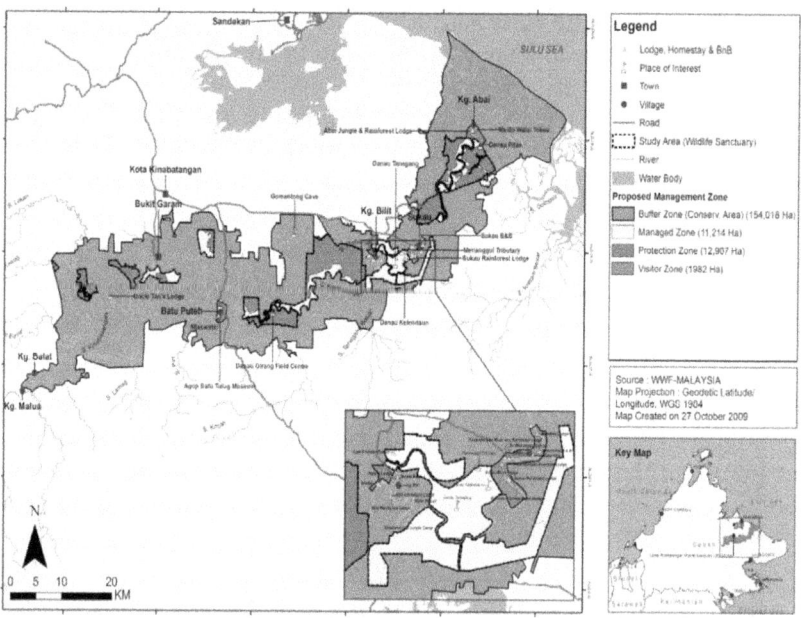

Figure 10.1 Location of Miso Walai Homestay

including the iconic orangutan, Pygmy elephants and proboscis monkeys, making it one of the top wildlife tourism destinations in Malaysia (Majail & Webber, 2006).

During the heydays of the timber industry in Sabah in the 1980s, the local *Orang Sungai* (River People) were mostly engaged by timber companies. They were well paid but the logging activities caused extensive damage to the ecosystems and wildlife habitats along the Kinabatangan floodplains. Under heavy international pressure, the Sabah government ceased large-scale and uncontrolled logging activities in the early 1990s and the Lower Kinabatangan was officially declared a wildlife sanctuary in 1996 (Lower Kinabatangan Wildlife Sanctuary or LKWS). However, the downside was the loss of jobs and income especially for those who used to work for the timber companies and as part time poachers. As a stop-gap measure to address the deteriorating economic problem, the government introduced cocoa farming, but the global collapse in the price of cocoa forced the local community to abandon this initiative (Payne, 1996). As a result, close to 90% of the local community were soon classified as hardcore poor which forced many of the local youth to leave their homes in order to seek jobs in Sandakan, Kota Kinabalu and Peninsular Malaysia (KOPEL Coordinator, interview, 2010).

To compensate for the loss of livelihood and income from the gazette of LKWS, a CBT project was initiated by the government, with the support of WWF Norway in 1997. The CBT project was given the acronym MESCOT (Model for Ecologically Sustainable Community Conservation and Tourism) and was given the task of building the capacity of the local community in Mukim Batu Puteh to operate a homestay programme linked to the ecotourism resources in the Lower Kinabatangan. In its formative years, MESCOT was administered by volunteers from the local youths in the village. In efforts to strengthen the homestay organization, the *Koperasi Pelancongan Mukim Batu Puteh* (KOPEL or tourism cooperative) was set up in 2003 under the auspices of the Cooperative Commission of Malaysia (SKM). Its membership increased incrementally from 72 in 2003 to more than 300 members in 2017, which is a testament of the success of the tourism cooperative.

The Never-Ending Conservation

The findings and conclusions presented in this paper are based on a longitudinal study carried out at Miso Walai Homestay from 2005 to 2017. In the initial stages of the longitudinal study (2005 to 2007), the researcher's contact with Miso Walai Homestay was limited to email correspondence and networking at tourism conferences during which the researcher became better acquainted with the coordinator of KOPEL. In 2008, the researcher included Miso Walai Homestay as one of the 10 case studies of CBT in the Asia Pacific region for a comprehensive study funded by the Asia Pacific Economic Cooperation (APEC). The findings and recommendations of this study were published in the Handbook on Community Based Tourism (Hamzah & Khalifah, 2009).

This marked the beginning of a long-term relationship between the researcher and Miso Walai Homestay, and also the second stage of the longitudinal research from 2008 to 2010, during which regular visits to conduct semi-interviews with the KOPEL management and members were carried out to collect baseline data, document the evolution of Miso Walai Homestay and identify the critical success factors. In the third stage (2010–2012), interviews were carried out with government agencies, tour operators and tourists as well as more in-depth interviews with the KOPEL management. The aim was to verify and refine the critical success factors. Throughout the longitudinal study, quantitative data such as the increase in tourist arrivals, income and other economic benefits were analysed using descriptive statistics while qualitative methods such as semi-structured interviews and focus group discussions were used to elicit the opinions of the key stakeholders. It should be highlighted that the researcher and his team stayed at a different homestay during each field survey to ensure that the interviews with the homestay operators were well represented. In the fourth stage of the longitudinal study (2013–2017) the researcher and the management of Miso Walai Homestay had developed a partnership by jointly organizing training programmes at the in-house training centre at Mukim Batu Putih. Through this association, the researcher was able to obtain a deeper insight into the dynamics and soul of Miso Walai Homestay.

Factors Contributing to Success of Miso Walai Homestay

While it is acknowledged that the success of Miso Walai Homestay was shaped by both exogenous and endogenous factors, this chapter will only focus on the latter. Essentially, exogenous factors include the growing popularity of the Kinabatangan area as a premier wildlife tourism destination, good accessibility from the tourism gateway town of Sandakan and the catalytic role of donors. However endogenous factors are deemed to be more appropriate in revealing the underlying motivation for the commitment of the local community towards responsible tourism. Based on the findings from the longitudinal study, six endogenous factors that had contributed to the success of Miso Walai Homestay as a CBT project are identified. They are outlined in the following subsections.

Ensuring 'buy in' from the local community and nurturing self-reliance

Most CBT projects are initiated by international donors and latterly embraced by governments given their populist nature (Hamzah &

Khalifah, 2009). Lately CBT projects initiated by international NGOs have been criticized for allegedly giving priority to the conservation agenda over local economic empowerment (Butcher, 2007; Ghasemi & Hamzah, 2010; Huxford, 2010), to the extent that the local community has often succumbed to the dependency trap (Harrison & Schipani, 2007). Initially the MESCOT project also suffered the same fate given that the 'local WWF staff were more concerned about their next project since the four-year funding from WWF Norway was about to end in 2001, instead of seeking fresh funding' (MESCOT Coordinator, interview, 2012). Nonetheless the former MESCOT Coordinator managed to secure extra funds from other donors and subsequently dedicated almost twenty years of his life in helping Miso Walai Homestay.

Given that most of the local community had predominantly been involved in an agrarian lifestyle all their lives, it was critical for them to be gradually exposed to the demands of tourism – which was then an alien phenomenon that they were about to encounter. To this end, the MESCOT organization implemented an incremental development approach by initially getting the whole community to be involved in the preparation of a tourism master plan followed by a capacity building programme that stretched for three years. The decision by MESCOT to spend three years focusing solely on capacity building was indeed risky as many KOPEL members interviewed recalled their initial apprehension with the CBT project which had yet to attract guests (interviews with KOPEL members, 2010, 2011, 2012). In fact, the former MESCOT Coordinator, who is an Australian, was even spat on by the villagers as he went from house to house trying to win over the local households.

Despite this, the Coordinator and his team of volunteers persevered and the early dissenters are now mostly active KOPEL members. Nonetheless such a lengthy capacity building process would have tested the patience of any rural community which explains why United Nations Development Programme (UNDP) recommended that 'quick-win projects' should be implemented along the way to pacify potential discontent among sections of the local community (Moeurn *et al.*, 2008). To overcome the initial apprehension from the local community, the MESCOT management turned to the senior citizens, which was an effective approach given that filial piety and respect for the elders were virtues that were and are still strong among the local community. Ultimately, leveraging on the respect for the elders was more effective that the house-to-house visits which triggered a gradual 'buy in' process among the local community. However, there was still a core group of initial dissenters, who were aghast at the thought of opening their homes to 'infidels *(kafirs)* and people who do not wash or speak their language' (interviews with local homestay operators, 2009, 2010, 2011). Most of the dissenters gradually softened their attitude and tolerated the early arrivals of guests to

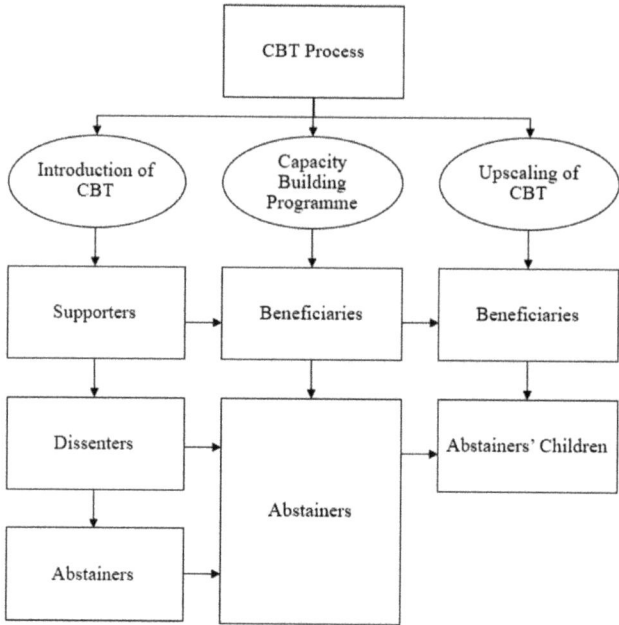

Figure 10.2 'Buy in' process of the local community

the homestays as long as they 'were not creating problems and were respectful of the local religion and culture'. In essence almost all the dissenters have mellowed to become abstainers, and although many of them have not joined KOPEL, they did not stop their children from being actively involved and remunerated through the CBT activities (interview with village youths, 2015) (see Figure 10.2).

Central to the lengthy capacity building programme was the nurturing of a self-reliant attitude among the local community and the rejection of outside aid even though many of them were still living in poverty during Miso Walai's formative years. In this light, the MESCOT management rejected financial aid from the government and channelled some of the 'philanthropic' income (Mitchell & Ashley, 2010) from international donors to finance interactive training approaches such as technical visits to tourism attractions within Sabah and neighbouring Sarawak. For many of the local community who went on the technical visits, it was the first time that they had ventured outside their village. Underlying these technical visits was the agenda that first-hand exposure to a vibrant tourism attraction would help the local community better understand what it takes to operate a successful tourism business (interview with KOPEL Coordinators, 2011).

During the post-mortem session after the technical visits, the local community was reminded that they should replicate the warmth and hospitality as well as the level of comfort, hygiene and safety that they had enjoyed during their visits, when hosting guests in their homestay. Needless to say, the technical visits were the most effective form of training that the three-year capacity building programme offered and no amount of classroom style teaching could match the effectiveness of the exposure trips (interview with homestay operators, 2010, 2012). It also signalled the broadening of the local community's worldview as they prepared themselves not only to enjoy the financial gains from the homestay programme but also the potential excitement of hosting guests from a different cultural and religious background (interview with homestay operators, 2010, 2011, 2012).

From local champion to broader-based organization

The role of a local champion(s) in CBT cannot be over emphasized, and even the transformation of an economically deprived Maori community in Kaikoura, New Zealand into one of the world's top whale watching tourism destinations owed its success to a local champion (Hamzah & Khalifah, 2009). Despite this acknowledgement, the concept of local champion or 'spark' according to Hatton (1999) is difficult to define given that they do not have to be a formal leader of a community. Recognizing this difficulty, Hamzah and Khalifah (2009) instead identified the attributes of a local champion which include trustworthiness, perseverance, selflessness, patience, courage and being visionary as well as having the 'ability to galvanize and transform the local community.

Adding to the fuzziness of the definition is the fact that the pioneer local champion at Miso Walai Homestay was an Australian tour guide turned WWF volunteer/MESCOT Coordinator. Uneasy with the way international tour operators were seemingly exploiting the local communities without a long-term commitment, the Australian volunteer decided to stay put at the village of Kampung Batu Puteh to develop and upscale its CBT project. As the first MESCOT Coordinator, he established an operational and financial system as well as training manuals to guide the capacity building programmes that he initiated (interview with KOPEL CEO, 2017). Furthermore, his extensive networking with international tour operators was instrumental in attracting tourists to Miso Walai Homestay. At the same time the first local youth to have completed tertiary education had just returned to the village to become a WWF/MESCOT volunteer. Recognizing his talent and potential, the MESCOT Coordinator handpicked the young graduate to be appointed as Assistant Coordinator, with him eventually replacing the Coordinator. The idealistic and yet practical young graduate stayed on to become the second local champion of Miso Walai Homestay until today.

Devoid of local youths with the capacity to assist him in managing MESCOT, the newly appointed Coordinator had no choice but to recruit the errant youths who had been working for the logging companies but had since lost their jobs. Financed by mainly international donors, MESCOT/KOPEL was already committed to supporting community-based reforestation programmes that were part of a larger scheme to reconnect the protected areas along the Kinabatangan River called the Kinabatangan Corridor of Life (partnership between the WWFMalaysia and the State government of Sabah). In this respect the errant youths had the advantage of having a sound knowledge of tree species and were used to working with outsiders – ideal qualities to lead MESCOT/KOPEL's reforestation activities notwithstanding their past history. With the training and counselling given to them by the young Coordinator, and their new responsibilities, the errant youths matured to become the second tier of local champions, who were not only competent in leading the reforestation activities but also became fully trained and certified nature guides. Several of the second-tier local champions were later offered good paying jobs by specialist tour operators or corporate social responsibility (CSR) companies operating in the Kinabatangan area to lead their own reforestation projects. Needless to say, the once errant youths are now committed joint custodians of the environment and champions of responsible tourism.

In addition to reforming the errant youths, the Coordinator identified and enticed back a core group of youths who were working in hotels in the major cities. Since leaving the village in search of stable jobs outside, they had acquired new skills, knowledge and a different worldview, and were brought back to focus on the hospitality aspect of KOPEL's operation. Initially they faced several challenges upon returning home such as the growing expectations from the local community and petty jealousies but the experience gained from their exposure to the competitive environment in the cities in Peninsular Malaysia had made them more resilient and able to readjust to the slower pace of life in the village.

As a CBT matures, however, it is argued that the local champion should make way for a broader-based organization, so as to curb potential manipulation and to ensure accountability and transparency (Moeurn et al., 2008). Currently, the management and operation of homestay programmes in Malaysia are mainly carried out by the Village Development and Security Committee (JKKK). However, previous research revealed that the JKKK's role in managing the homestay programme had been ineffective and susceptible to manipulation and cronyism (ECERDC, 2009). In this light, the setting up of KOPEL in 2003 was able to separate the commercial aspects of the homestay operation from being entangled by the other mundane and practical duties of the JKKK.

In addition, the establishment of KOPEL brought about a new dimension to the CBT organization in terms of commercial outlook and professionalism. To ensure continuity, the pioneer batch of MESCOT volunteers were mainly absorbed as full-time KOPEL staff to manage the five bureaus of the tourism cooperative. The second local champion is currently the Chairman of KOPEL, who has since delegated most of the decision-making responsibilities to the second tier of local champions, under the leadership of an Executive Manager who was at one time cutting down

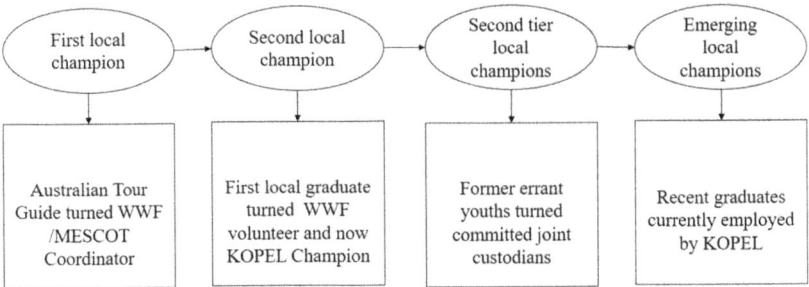

Figure 10.3 Evolution of champions at Miso Walai Homestay

trees for timber companies until the early 1990s (interview with KOPEL Chairman, 2014) (see Figure 10.3). Starting from 2015, several children of the KOPEL EXCO and members had returned to the village upon completing tertiary education in the cities. At least five of them have been appointed into managerial positions in the KOPEL management. While this has significantly improved KOPEL's professionalism, especially in terms of financial management, it has also caused resentment among the pioneer volunteers turned KOPEL managers, who are being paid less than the recent graduates (KOPEL manager, pers. comm., 2017).

Fostering a sense of ownership and restoring pride in the local community

During the formative years of MESCOT, the local community's spirit was at its lowest ebb given that many had lost their lucrative jobs for the timber companies to the extent that the out-migration of youths to find jobs in the cities became prevalent (Payne, 1996). To address the situation, MESCOT (and subsequently KOPEL) decided to create a community project that would attempt to heal and restore the local community's self-esteem and pride (interview with KOPEL Chairman, 2009). Towards this end, KOPEL submitted a proposal to the Sabah Forest Department (SFD) in 2002 to construct the Tungog Rainforest Eco Camp (TREC) in the adjacent forest reserve. Either due to luck or good timing, the application coincided with SFD's new approach in protected area management which was shifting towards a more collaborative partnership with local communities that was forced upon by the Sabah government's commitment to the Green Economy (interview with SFD officer, 2010). In what could be viewed as a watershed decision, SFD not only gave permission but also provided a fund of just under RM900,000 (USD 225, 000) for the construction of an ecolodge type accommodation to be managed by KOPEL, albeit following stringent standards. Instead of taking the easy way out by engaging a professional contractor to construct TREC, KOPEL decided that it should be a local community project and the ideal opportunity to

Figure 10.4 Tungog Rainforest Eco Camp

encourage 'buy in' and foster a strong sense of ownership among the local community. With extra funding from international donors, TREC was constructed by the local community with the help of voluntourists. Those who could not contribute financially or physically in the construction work donated building materials such as the much prized *belian* wood (*Eusideroxylon zwageri*) that they have been keeping in their compound for years. Although it took almost five years for TREC to be completed, the 'journey' that the local community went through in partnership with voluntourists succeeded in restoring their pride and most importantly, nurtured a sense of local ownership of the project, both physically and spiritually. Today TREC is revered as a symbol of the local community's spirit, determination and grit to overcome adversity (Figure 10.4).

In 2016 KOPEL members made the collective decision to refurbish the ecolodges at TREC to improve the level of comfort of the guest facilities without compromising the principles of environmental sustainability and Green Design. By doing so, TREC has become one of the top attractions in the Kinabatangan especially for wildlife watching in a pristine forest environment. Furthermore, TREC has been able to attract a continuous stream of international students in the past two years since its refurbishment to conduct research and field experiments (interview with KOPEL's Activity Manager, 2017). This low-impact form of educational tourism has further strengthened Miso Walai Homestay's growing reputation as a responsible tourism destination.

The local community's pride in TREC, however, has been severely tested over the past few years. Fronting the ecolodges at TREC lie Tungog

Lake, which is an ox bow lake that used to be the habitat for a variety of freshwater fishes and birds. Unfortunately, the lake has been invaded by an invasive species of weeds called the *Salvina molesta* that originates from South America. Despite conducting regular clean-up of the lake through community-based efforts with the help of volunteers, the weeds kept coming back and are becoming a serious threat to the ecosystem and ecological services of the lake, besides affecting the aesthetics of TREC. While the KOPEL management is still searching for a long-term biological solution for this menacing threat, the local KOPEL members had approved a significant amount of money from the Community Fund to be used for clearing the weeds despite the fact that previous successes had only been short lived. The monetary sacrifice that the local community demonstrated is a clear reflection of their sense of ownership and determination in safeguarding TREC and its surrounding ecosystems and a collective symbol of pride (interview with KOPEL CEO, 2017).

Sustaining commercial viability and developing career paths

Miso Walai Homestay is among the top three homestays in Malaysia in terms of tourist arrivals and receipts (Tourism Malaysia, 2018). Tourist arrivals increased by 19.5% from 3087 tourists in 2016 to 3689 tourists in 2017. In terms of receipts, KOPEL generated an income of around RM 1.8m (USD 460,000) in 2017, which represented an increase of 10% from the previous year's income of almost RM 1.7m (USD 420,000) (KOPEL, 2018) (see Table 10.1).

The increase in tourist arrivals and income was achieved by increasing the breadth and depth of the homestay experience. In the early days of Miso Walai Homestay, staying with an adoptive family and participating in the local way of life was the highlight of the homestay experience, which explains why the homestay operators were initially the highest income earners. It is interesting to note that the boatmen and local guides have since overtaken the homestay operators as the highest income earners, with the majority earning an average income of more than RM3000 (USD1000) per month during the peak season. Furthermore, seasonality is no longer an issue with tourists arriving all year round since 2016 thus ensuring a stable income for KOPEL members. Besides income from the homestay and

Table 10.1 Tourist arrivals and income received by KOPEL, 2015–2017

Year	Tourist arrivals	Growth %	Income	Growth %
2017	3689	19.5%	RM 1,844,067	9.9%
2016	3087	16.1%	RM 1,677,751	13.9%
2015	2660	−5.5%	RM 1,473,559	−20.0%

Source: KOPEL (2018).

boat/guiding services, the local community are also paid to perform other tasks such as cultural performances, food catering, local transport and the production of local handicrafts. In addition, members are allowed and encouraged to set up family cooperatives under the umbrella of KOPEL, for instance to develop fish farms and biomass production, etc.

Towards the end of 2011, KOPEL carried out a survey of tourist preferences and satisfaction levels which revealed that Miso Walai Homestay 'lost' around 1700 potential guest nights in that year because guests had no option but to stay in the registered homestays. The survey also revealed that guests would have stayed extra nights if they were other options available although most of them had enjoyed the experience of staying one night with an adoptive family (interview with KOPEL Chairman, 2013). In response to this survey finding, the KOPEL management decided to create other forms of accommodation to cater for the needs of different market segments. For instance, a hammock tent site was developed which became the most popular accommodation facility in Miso Walai Homestay, especially with student groups (see Table 10.2). After several years of relative neglect, the ecolodges in TREC were given a new lease of life through a substantial refurbishment exercise to attract high value guests who were looking for a unique experience of staying in a forest so as to be in direct contact with nature and wildlife. Expanding the breadth of the accommodation facilities resulted in an increase in the average length of stay (ALOS) from two nights in 2011 to 2.5 nights in 2012, and a corresponding increase of 26% in terms of income.

In 2014 the Maandastay (villagestay) stand-alone accommodation was opened to the public, offering eight twin sharing rooms and two dormitories that could accommodate a total of 50 beds. In contrast to the traditional homestay, Maandastay offers a unique experience of staying in a village environment but in almost total privacy, which makes it popular with domestic tourists, businessmen and conference/training groups (KOPEL, 2018). It is worth noting that the standalone chalets have now surpassed the other variants to become the most popular accommodation in 2018 (interview with KOPEL CEO, 2019). However, the development of

Table 10.2 Types of homestay variants at Miso Walai Homestay

Variants	Privacy	Intimacy with hosts	Authentic rural experience	Market segment
Traditional homestay	Low	High	High	Students FITs Voluntourists
Adventure Camp	Low	Low	Low to medium	FITs Student groups
Hammock camp	Medium	Low	Low to medium	Student groups
Maan da stay (village stay)	High	Medium	Low to medium	Conference groups Small families Businessmen

Maandastay has also caused resentment among the registered homestay operators who complained that potential guests were shunning away from staying with them in preference for the privacy being offered by the stand-alone accommodation. In the typical 'decision-making by consensus' style of management practised by KOPEL, the homestay operators were consulted and subsequently given soft loans to improve the physical condition of their homes and guest amenities as well as refresher courses (interview with KOPEL Chairman, 2016). To surmise, the expansion in the breadth of tourist accommodation at Miso Walai Homestay has enhanced the competitiveness of Miso Walai Homestay and expanded its market to ensure its long-term commercial viability but poses a new challenge in terms of ensuring an equitable distribution of income among its members.

By the same token, KOPEL has taken effective action in increasing the depth of the tourist experience which has also contributed to an increase in the length of stay and tourist spending. In this light, the local guides were sponsored to attend advanced training and certification so as to improve their story telling and product knowledge while new activities were added such as a boat ride cum trekking package to an archaeological site (interview with KOPEL CEO, 2016). As a tourism cooperative, KOPEL only keeps 3% of the income it generates to pay the salary of its staff, and the Cooperative Fund could also be used as a source of micro credit for its members. The bulk of KOPEL's income, however, goes back to its members in terms of shares and for services rendered. In addition, KOPEL allocates a special 'compassion' fund to be disbursed to its members during bereavements, weddings and as a token contribution when a local youth enters university. KOPEL's revolving fund is also transparent in its operations given that the annual financial report is audited and scrutinized by the Cooperative Commission of Malaysia (SKM). To ensure better transparency and accountability, KOPEL started to engage the services of a certified auditing firm to prepare its financial report since 2011 (interview with KOPEL Treasurer, 2015).

Equally important as a commercial viability is a secure career path for youths entering the tourism industry. For community-driven tourism projects in developed economies such as the Kaikoura Whale Watch in New Zealand, it was possible for a local tour guide to become a boat captain and eventually the CEO along a clear career path (Hamzah & Khalifah, 2009). Towards this end, the willingness of the management in sponsoring the training cost of its young staff members is crucial. For KOPEL young members, their lack of formal education is a handicap towards their progression along a desired career path. Nevertheless, KOPEL operates a meritocracy system in which promotions are based on talent, performance and attitude over paper qualification. The promotion of several long-term volunteers, who are without paper qualification, to key managerial positions is a testimony to the commitment of KOPEL in creating a secure career path for its members. Nonetheless as the younger generation becomes better educated, and fuelled by a desire to give back to the community, KOPEL is under pressure to accommodate fresh graduates among the sons and daughters of its members as part of its management without upsetting the pioneer members who had made huge sacrifices for Miso Walai Homestay. Suffice to say that having a clear career path is regarded by the local community as being more important than merely economic benefits (interview with KOPEL manager, 2017).

Using philanthropic flows to enhance local capacity

The term philanthropic flow is used by Mitchell and Ashley (2010) to represent income earned by CBT projects in the form of donations, CSR programmes and volunteerism activities. Given that most CBTs are initiated by non-governmental organizations (NGOs), philanthropic flows are usually allocated for conservation-related activities such as tree planting and ecosystem restoration although ecotourism development is often included as part of the overall aid package. Since the establishment of MESCOT in 1997, Miso Walai Homestay had received almost RM4m (USD1m) in philanthropic flows from international donors such as ARCUS Foundation and American Forests, and lately from Sabah government agencies, mainly to finance reforestation projects.

Over the years, KOPEL has been able to use the philanthropic flows that it received for reforestation projects to create around 80 jobs for the local women. Each month, KOPEL paid out a total salary of between RM28,000 (USD9000) to RM80,000 (USD28,000) to these women which helped supplement the household income of their families (interview with KOPEL Chairman, 2010). More importantly KOPEL had used the opportunity created by donor funding to enhance the traditional ecological knowledge and skills of its members through their active participation in the reforestation projects instead of merely using philanthropic flows to fulfil the conservation agenda stipulated by the donors.

In what could be regarded as a game-changing action, KOPEL submitted a detailed proposal to SFD in 1999 to carry out community reforestation projects for the agency (interview with KOPEL Chairman, 2010). As a testimony to their growing reputation, and to their pleasant surprise, KOPEL was awarded a RM1.3m. (USD320,000) contract by SFD in the same year to carry out a pilot community-based reforestation along the Kinabatangan River as part of the Corridor of Life project. By 2014 KOPEL had replanted more than 332, 171 trees of 23 different species covering an area of 225 ha. In addition, KOPEL had supplied more than 15,000 seedlings to other reforestation programmes in LKWS. Instead of completing the reforestation projects merely to honour their contract with the donor agencies, KOPEL was able to build the capacity of its members to become competent contractors with the skills, ability and confidence to bid for reforestation project tenders, which they have managed to secure at least until 2019 (interview with KOPEL Chairman, 2016). Furthermore, KOPEL was appointed as the contractor for the EU REDD+ reforestation project covering the Kinabatangan area in 2017.

Towards this end, the reforestation projects were successful in re-creating habitats for wildlife along the Kinabatangan River which dovetailed with the wildlife tourism activities being offered by KOPEL. Tree planting has also become a major activity for tourists staying at Miso Walai Homestay but the activity is not limited to 'environmental tokenism' in the form of a 'plant a tree' ceremony that most of the surrounding ecolodges are offering to tourists (Ministry of Tourism, Arts and Culture Malaysia, 2017). Instead voluntourists spend up to four weeks or more being involved in systematic reforestation which covers the whole spectrum of silviculture to tree maintenance (interview with KOPEL Reforestation Manager, 2015).

In 2012 the Ministry of Rural and Regional Development created a nationwide *Desa Lestari* (Sustainable Villages) programme in which Miso Walai Homestay was nominated and shortlisted as one of the contenders for the prestigious grant that was initially limited to only four performing villages. Fearing that receiving a substantial grant from the government might be going against the local community's anti-handout mentality, KOPEL took some time to engage and consult its members before going ahead with the nomination. The main reason for this decision was that KOPEL needed to upscale its tourism offerings and to reach out to a wider market so as to generate enough income to sustain its commercial viability and safeguard the well-being of its ever-increasing members (interview with KOPEL Chairman, 2013). After a lengthy and stringent process, Miso Walai Homestay was selected as one of the winners that came with a RM5m. (USD 1.3m) allocation for the construction of a training centre cum office, Maandastay stand-alone accommodation, a new jetty, restaurant, six fibreglass boats and four vans (interview with KOPEL Chairman, 2015). The physical development was completed and became operational in 2014. This development, which has enhanced the appeal and image of Miso Walai Homestay, has in turn translated into an annual increase of more than 10% in tourist arrivals and receipts since 2015 (refer to Table 10.1).

It is not uncommon for homestays in Malaysia to be given a community centre to conduct meetings and training related to CBT under the Malaysian Homestay Experience budget. However, many of these facilities became underutilized and end up as 'white elephants' (MOTAC, 2017). If indeed there was a risk that the new training centre at Miso Walai Homestay would suffer the same fate, the initial trepidation was soon dispelled when KOPEL was able to conduct a series of workshops and training as well as host study visits at the centre. The icing on the cake was achieved when KOPEL was appointed as the coordinator of the Coaching and Mentoring Programme funded by the Cooperative Commission of Malaysia (SKM) in 2017 (SKM, 2017). In relation to this, a pilot programme was initiated in 2017 which involved a one-year

Figure 10.5 New training centre as part of *Desa Lestari* programme

mentoring programme for tourism cooperatives in Sabah and Sarawak; this programme used the KOPEL training centre as a base for hands-on training and the surrounding activities and attractions as the 'outdoor laboratory' (Figure 10.5).

As part of the mentoring programme, the researcher was appointed by SKM to be the coach or master trainer, which provided him with the ideal opportunity to get a deeper insight into the 'soul' of Miso Walai Homestay. Central to the content of the mentoring programme is the role of the KOPEL management and youths as mentors and trainers. Besides generating extra income, this form of philanthropic revenue also empowered KOPEL as a training provider as well as raised the profile and self-esteem of its members (interview with KOPEL youth leaders, 2017). Judging from the endless stream of technical visits from government agencies and CBT projects across the country, Miso Walai Homestay is gradually developing into a model for responsible tourism, which was triggered by the philanthropic flows it managed to attract since the late 1990s.

Establishing early partnerships with specialist tour operators

One of the inherent weaknesses of CBT projects is its lack of contact with the outside world given its frequently rural, remote and sometimes isolated location. Despite their locational disadvantage, remote CBT projects have been able to flourish through capable leadership, good management and a strong community spirit (MOTAC, 2017). However, upscaling a remote CBT along the value chain is challenging without the benefit of being connected to a network of tour operators for help to secure new markets (Hamzah & Khalifah, 2009). Such partnerships will pave the way for CBT to gradually become a mainstream tourism product, rather than remaining as a niche attraction. In seeking partners in the tourism industry, timing is crucial. Establishing partnerships at the onset of a CBT project will boost its comparative advantage of being part of an international or national tourism circuit. However, a CBT that is lacking in local capacity during its inception might end up being engulfed by opportunistic tour operators if it were to follow this path. On the other hand, a CBT that is not ready to set up partnerships until it achieves maturity and a certain level of success, might be overtaken by other CBT projects and competitors (Hernandez-Maskivker *et al.*, 2018).

Miso Walai Homestay was fortunate because, prior to joining MESCOT, its first coordinator worked as a tour guide for specialist tour operators in Sandakan and the Kinabatangan area. This made it relatively easy for him to set up partnerships with voluntourism and specialist tour operators such as Gecko, Borneo Eco Tours, Intrepid Travel, Exodus, Raleigh International and Global Vision International (GVI). Through these effective partnerships, Miso Walai Homestay recorded a seven-fold increase in tourist arrivals, from 1200 night-stays during the inception of MESCOT in1997 to 7960 night-stays in 2010.

As in most other CBT projects, Miso Walai Homestay set up partnerships with tour operators primarily to overcome their lack of capacity in carrying out extensive marketing and promotion. Additionally, collaborating with voluntourism companies created the opportunity for MESCOT to seek volunteers and additional funding to develop its flagship community project, the Tungog Rainforest Eco Camp (TREC) in 2002. Voluntourism activities tend to be *ad hoc* in nature, often resulting in frustration on the part of the voluntourists, who expect a more active involvement of the local community in designing the activities (Synman, 2014). In the case of TREC, the five years KOPEL took to complete the construction and another two years before it was operational provided a focus for both the volunteers and the local community to work together for a common cause.

Finally, establishing partnerships with international operators at the onset of the CBT project had set a high standard for the local community to attain due to their lack of capacity and exposure to tourism. By the same token, raising the bar demanded a colossal effort from the impoverished local community, which explains why MESCOT had to devote three years solely for capacity building before Miso Walai Homestay was ready to receive guests. Safety and security were crucial to the overall tourist experience demanded by the tour operators; therefore, risk assessment and management were incorporated into the training modules. In addition, the local guides were trained to conduct safety briefings before each outdoor activity as well as ensure that safety procedures were diligently adhered to during such activities. Over the years, other forms of safety requirements were demanded by the tour operators such as a rescue and evacuation protocol which was fulfilled by KOPEL (interview with KOPEL CEO, 2016).

Given the experiential nature of the homestay programme, religious and cultural sensitivities have to be observed, and potential conflicts and misunderstanding were minimized by developing a code of conduct for guests, which was disseminated during check-in and the 'handing over' session. This includes a dress code, a list of etiquette and a reminder to respect the local religion and culture (interview with KOPEL Chairman, 2010). Despite the early difficulties that the local community had to go through to achieve the high standard set by the international operators, the intensive capacity building programme that was implemented made it seamless for KOPEL to move up the value chain as it matured.

Discussion and Conclusion

According to a Japanese professor who accompanied the researcher during his field survey in 2013, the success of Miso Walai Homestay was 'a miracle' and a 'one-in-a-million success story'. Likewise, many of the participants of training workshops held there were amazed that a previously impoverished community could find the inner strength to lift their spirits and overcome adversity, and yet remain humble and fully committed to an agenda that they set 20 years ago. The success of Miso Walai Homestay seems more remarkable in comparison with the other three CBTs along the Kinabatangan River, who are still struggling with internal rivalry, lack of strong leadership and being caught in the dependency trap (SKM Coaching and Mentoring Programme, 2017). Belonging to the same *Orang Sungai* ethnic group and enjoying the same exogenous factors such as good accessibility and the iconic image of the Kinabatangan area as a wildlife tourism destination, the other CBTs, however, did not go through a systematic and continuous capacity building process to prepare their communities.

In recognition of its past achievements, Miso Walai Homestay had won numerous international and national awards. What is seldom recognized though is its contribution towards the creation of the Kinabatangan Corridor of Life through its community reforestation projects. Towards this end, the process of replanting the riparian reserves had significantly contributed to the growth in voluntourism while the re-created wildlife habitats are crucial for the sustainability of non-consumptive wildlife tourism in the Kinabatangan. Nonetheless, it has to be accepted that KOPEL's noble community efforts are taking place in an area that is experiencing serious environmental threats from oil palm plantation owners, which has resulted in the deterioration of the Kinabatangan River and conflicts with wildlife, especially the iconic Pygmy elephants. Unless there is a full commitment by all stakeholders to protect the ecosystems along the Kinabatangan floodplain and to hasten the completion of the Corridor of Life, Miso Walai Homestay is likely to lose its comparative advantage.

During the infamous incursion of terrorists from the Philippines into Lahat Datu town in 2013, Miso Walai Homestay suffered a substantial drop in tourist arrivals and income due to booking cancellations. Although Lahad Datu is located by the coast and about a 45-minute drive from Miso Walai Homestay, the travel advisories issued by countries such as the UK,

Australia and Japan were the main reason for the cancellations which went on for two years. For the first time in years, KOPEL's resilience was severely tested given that the alarming decrease in tourist arrivals was threatening the livelihood and morale of its members. In response to this predicament, KOPEL took desperate measures upon consulting its members such as approaching tourists at the tourism attractions in the tourism gateway town of Sandakan and promoting Miso Walai Homestay to tourists from Peninsular Malaysia. There were two lessons learned from the mini crisis (interview with KOPEL Chairman, 2014). First is not to put all the eggs in one basket which explains why KOPEL is currently reaching out to different market segments so as not to be over reliant on voluntarists and free independent travelers (FIT) brought in by specialist tour operators. Although tourist arrivals have stabilized and are no longer affected by seasonality, KOPEL is currently exploring new avenues for income diversification such as biomass production (from oil palm waste), aquaculture and as training providers (interview with KOPEL CEO, 2017). Second is to ensure that KOPEL's financial system is able to withstand and recover from unexpected shocks such as the repercussions of the Lahad Datu incursion. On hindsight, the KOPEL CEO reflected that developing a sound financial model and system should be the first priority during the setting up of a CBT, and this strong message was stressed during the Coaching and Mentoring workshops (interview with KOPEL CEO, 2017).

Miso Walai Homestay is at a crossroads. The increase in tourist arrivals is no longer confined to voluntourists and FITs who are compatible with the ideals of responsible tourism. Over the years, young Middle Eastern student groups, international schools, domestic small groups and families and incentive groups have discovered Miso Walai Homestay. Consequently, there is the likelihood that the existing host–guest relationship, especially mutual respect and learning from each other that the local community had been accustomed to, will be challenged.

As proud parents within the community welcome the return of sons and daughters with paper qualification and the idealistic notion of giving back to the community, there is hope that the effectiveness of the CBT in tackling rural precarity (Rigg *et al.*, 2016) could be demonstrated in Miso Walai Homestay. Conversely, accommodating these fresh graduates by absorbing them and their much-needed skills-set in KOPEL's management would put a strain on its financial resources besides triggering petty jealousies. In this light, expanding the economic base and ensuring an equitable distribution of a relatively small pie is critical. Towards this end the much admired and role model success of Miso Walai Homestay should only be viewed as a stepping stone and not a panacea for rural underdevelopment issues in Malaysia. Nonetheless, the local community's triumph over adversity and their proven resilience will put them in good stead as the next generation reinvents Miso Walai Homestay.

In the wider context, the success of Miso Walai Homestay could inspire the local communities and other stakeholders along the Kinabatangan Corridor of Life to increase their commitment and efforts. The reconnection of the fragmented protected areas and community conserved areas along this ecological corridor goes beyond biodiversity conservation. Its success will achieve a higher goal of reinventing the harmonious relationship between communities and nature, that is distinctly Asian in philosophy – with responsible rural tourism playing an essential role.

References

Butcher, J. (2007) *Ecotourism, NGOs and Development*. Abingdon: Routledge.
Chettiparamb, A. and Kokkranikal, J. (2012) Responsible tourism and sustainability: The case of Kumarakom in Kerala, India. *Journal of Policy Research in Tourism, Leisure and Events* 4 (3), 302–326.
East Coast Economic Region Development (ECERD) (2009) East Coast Economic Region Master Plan. Kuala Lumpur: ECERDC.
Ghasemi, M. and Hamzah, A. (2010) The use of Delphi technique to determine variables for the assessment of community-based ecotourism. *Proceedings of Regional Conference on Tourism Research* (p. 89). Penang: Social Transformation Platform, Universiti Sains Malaysia.
Goodwin, H. (n.d.) Responsible tourism. http//www.haroldgoodwin.info/responsible-tourism/
Goodwin, H. (2012, April 17). There is a difference between sustainable and responsible tourism. *Harold Goodwin: Taking Responsibility for Tourism*. See http://haroldgoodwin.info/there-is-a-difference-between-sustainable-and-responsible-tourism/ (accessed 17 April 2012).
Hamzah, A. and Khalifah, Z. (2009) *Handbook for Community-Based Tourism: How to Develop and Sustain CBT*. Singapore: Asia Pacific Economic Cooperation.
Harrison, D. and Schipani, S. (2007) Lao tourism and poverty alleviation: Community-based tourism and the private sector. *Current Issues in Tourism* 10 (2), 194–230.
Hatton, M.J. (1999) *Community-based Tourism in the Asia-Pacific*. Singapore: Asia-Pacific Economic Cooperation.
Hernandez-Maskivker, G., Lapointe, D. and Auino, R. (2018) The impact of volunteer tourism on local communities: A managerial perspective. *International Journal of Tourism Research* 20 (5), 650–659.
Hjalager, A.-M. (1996) Tourism and the environment: The innovation connection. *Journal of Sustainable Tourism* 4 (4), 201–218.
Homestay Operators (2009) (A. Hamzah, Interviewer)
Homestay Operators (2010) (A. Hamzah, Interviewer)
Homestay Operators (2011) (A. Hamzah, Interviewer)
Homestay Operators (2012) (A. Hamzah, Interviewer)
Huxford, K.M. (2010) *Tracing Tourism Translation: Opening the Black Box of Development Assistance in Community-Based Tourism in Vietnam*. Christchurch, New Zealand: University of Canterbury.
KOPEL Activity Manager (2017) (A. Hamzah, Interviewer)
KOPEL CEO (2016) (A. Hamzah, Interviewer)
KOPEL CEO (2017) (A. Hamzah, Interviewer)
KOPEL CEO (2019) (A. Hamzah, Interviewer)
KOPEL Chairman (2009) (A. Hamzah, Interviewer)
KOPEL Chairman (2010) (A. Hamzah, Interviewer)
KOPEL Chairman (2013) (A. Hamzah, Interviewer)

KOPEL Chairman (2014) (A. Hamzah, Interviewer)
KOPEL Chairman (2015) (A. Hamzah, Interviewer)
KOPEL Chairman (2016) (A. Hamzah, Interviewer)
KOPEL Coordinator (2010) (A. Hamzah, Interviewer)
KOPEL Coordinators (2011) (A. Hamzah, Interviewer)
KOPEL Manager (2017) (A. Hamzah, Interviewer)
KOPEL Member (2010) (A. Hamzah, Interviewer)
KOPEL Member (2011) (A. Hamzah, Interviewer)
KOPEL Member (2012) (A. Hamzah, Interviewer)
KOPEL Reforestation Manager (2015) (A. Hamzah, Interviewer)
KOPEL Treasurer (2015) (A. Hamzah, Interviewer)
KOPEL Youth Leaders (2017) (A. Hamzah, Interviewer)
KOPEL (2018) *KOPEL Annual Report 2017*. Sandakan: KOPEL.
Leslie, D. (2012) Introduction. In D. Leslie (ed.) *Responsible Tourism. Concepts, Theory and Practice* (pp. 1–16). Wallington: CABI.
MacDonald, R. and Jolliffe, L. (2003) Cultural rural tourism: Evidence from Canada. *Annals of Tourism Research* 30 (2), 307–322.
Majail, J. and Webber, D.A. (2006) *Human Dimension in Conservation Works in the Lower Kinabatangan: Sharing PFW's Experience*. Fourth Sabah-Sarawak Environmental Convention. 5 September 2006, Kota Kinabalu, Sabah, Malaysia.
MESCOT Coordinator (2012) (A. Hamzah, Interviewer)
Mihalic, T. (2016) Sustainable-responsible tourism discourse – towards 'responsustable' tourism. *Journal of Cleaner Production* 111, 461–470.
Ministry of Tourism, Arts and Culture Malaysia (MOTAC) (2017) *Business Strategies for Upscaling the Malaysia Homestay Experience*. Putrajaya: MOTAC.
Mitchell, J. and Ashley, C. (2010) *Tourism and Poverty Reduction: Pathways to Prosperity*. London: Earthscan.
Moeurn, V., Khim, L. and Sovanny, C. (2008) Good practice in Chambok community-based ecotourism project in Cambodia. In P. Steele, N. Fernando and M. Weddikkara (eds) *Poverty Reduction that Works: Experience of Scaling up Development Success* (pp. 3–19). London: Earthscan.
Payne, J. (1996) *The Kinabatangan Floodplain: An Introduction*. Kuala Lumpur: WWF Malaysia.
Rigg, J., Oven, K.J., Basyal, G.K. and Lamichhane, R. (2016) Between a rock and a hard place: Vulnerability and precarity in rural Nepal. *Geoforum* 63–74.
Sabah Forestry Department (2010) (A. Hamzah, Interviewer)
Scheyvens, R. (2002) *Tourism for Development: Empowering Communities*. Michigan: Prentice Hall.
Sharpley, R. (2002) The challenges of economic diversification through tourism: The case of Abu Dhabi. *International Journal of Tourism Research* 4 (3), 221–235.
SKM (2017) Coaching and Mentoring Programme.
Synman, S. (2014) Assessment of the main factors impacting community members' attitudes towards tourism and protected areas in six southern African countries. *African Protected Area, Conservation and Science* 56 (2), 1–12.
Tooman, L.A. (1997) Applications of the life-cycle model in tourism. *Annals of Tourism Research* 24 (1), 214–234.
Tosun, C. (2000) Limits to community participation in the tourism development process in developing countries. *Tourism Management* 21 (6), 613–633.
Tourism Malaysia (2018) *Tourism Fast Facts 2017*. Putrajaya: Tourism Malaysia.
Village Youths (2015) (A. Hamzah, Interviewer).

11 'MlupBaitong' – A Pioneer in Responsible Rural Tourism in Cambodia

Trevor H.B. Sofield

Introduction: Cambodia – Geography and Economy

Before assessing the contribution of MlupBaitong (MB) to responsible rural tourism development, a brief overview of Cambodia's geography and economy is presented, followed by a short analysis of the government's tourism policy on environment. Cambodia is a tropical country situated in South-east Asia that borders Thailand in the West, Laos in the North and Vietnam in the East. It has 440 km of southern coastline with 48 small offshore islands in the Gulf of Thailand. The country covers 181,000 sq km and is mostly flat with low mountains in the North and South-east. It is intersected north–south by the Mekong River which drains into the delta of southern Vietnam. The river also feeds South-east Asia's largest lake in Cambodia's interior, Tonle Sap, which in the wet season with the annual flooding of the Mekong expands from 4500 sq km to more than 16,000 sq km and is Asia's most productive freshwater fishery.

Cambodia's population of 16.2 million (2016 estimate) is 97% ethnically Khmer and Buddhist, 50% of whom are under the age of 25 years. About 75% are rural, 25% live in urban areas and Phnom Penh, the capital city, has a population approaching 2 million as the rural–urban drift accelerates to about 5% p.a. There is a huge disparity between the rich and the poor with the World Bank (2017) reporting that 2% of the population owned more than 70% of the country's entire resources, with about 15% living below the poverty line and another 40% living on or minimally above the poverty line (World Bank, 2017). Cambodia is one of the poorest countries in Asia and long-term economic development confronts a challenging future as it is 'inhibited by endemic corruption, limited human resources, high income inequality, and poor job prospects' (US Government World Fact Report, July 2018).

From a low baseline, Cambodia's gross domestic product (GDP) has grown at an annual rate of 7% since 2011, and this has elevated it from one of the world's 25 least developed countries (LDCs) to the status of a lower middle-income country (World Bank Group, 2016) GDP (at purchasing power parity) in 2017 was estimated at USD64.21 billion and Cambodia was ranked 105th globally. The tourism sector is the largest contributor to national growth, employing more than 500,000 people. Other prominent sectors graded according to importance are the garment and footwear industry (employing more than 700,000 people, 85% of whom are women), construction and real estate, and primary industry (cash crops are rice, rubber, corn, vegetables, cashews, cassava (manioc, tapioca) and silk). With the exception of some rice, vegetables and cassava, the other agricultural crops are dominated by large properties rather than small family holdings. Rural communities comprise most of those living at a subsistence level below the poverty line. These communities lack basic infrastructure (roads, electricity, water, sewage, clinics, schools) with low levels of education and productive skills.

Cambodian Government Tourism Policy

Cambodia is not noted for its rural and ecotourism ventures, although its natural resources are abundant and similar to those of its neighbours, Thailand, Laos and Vietnam. It still has extensive areas of relatively untouched forests with high biodiversity in the north-east mountains bordering Vietnam and Laos, and in the Cardamom Ranges (more than 1 million hectares) in the south-east bordering Thailand. While the Ministry of Tourism over the past decade has established a policy framework that provides a sound regulatory foundation for responsible rural and nature-based tourism development, in reality a number of factors have militated against implementation and enforcement. Cambodia has been riven by internal warfare since the 1970s, suffering the Khmer Rouge revolution and accompanying genocide from 1976–1979, occupation by Vietnam until 1991 and civil war until 1998. The quality and conservation of much of its wildlife and wilderness have been due to this unrest and inaccessibility rather than any intentional policies of protection and preservation (Sofield & McTaggart, 2005; Sofield, 2009; Reimer & Walter, 2013).

Peace has been characterized by rapid exploitation of natural resources with little or no accountability for environmental concerns, and land grabbing continues on a significant scale by the rich and powerful (the Khmer Rouge destroyed virtually all land records, so titles are mostly non-existent). These factors have been compounded by rampant corruption that undercuts the legitimacy and functioning of the state so that the rule of law becomes tenuous, and the capacity of the public service (which is also complicit in corruption) to implement and monitor rules and regulations is emasculated. Enforcement is largely non-existent, and even designated protected areas face degradation from frequent illegal logging, hunting and wildlife trade, mining, illegal settlement and land development (Reimer & Walter, 2013). In a similar vein, Schipani (2017: 19) noted that 'despite well-designed management plans, the weak enforcement of laws and regulations, development of extractive industries, expanding commercial agriculture, and wildlife poaching (are) key threats to tourism development in and around the country's natural protected areas'. In this context, in 2016, the Transparency Index rated Cambodia 156th worst for corruption out of 176 countries, and the World Bank Governance Indicators Project (2017) ranked corruption in Cambodia as 17th worst out of 228 countries (i.e. no. 211). Since rural community-based tourism (CBT) in Cambodia is small scale, it has been of little interest to the central Government despite its paper commitment to the UN Sustainable Development Goals 2030 (especially Goal No.1, elimination of poverty) and it has left it to aid donors and international/non-governmental organizations (I/NGOs) (of which there are 2200 registered in the country) to take the lead.

As noted, the Ministry of Tourism struggles to expand its work in supporting responsible rural tourism and CBT because the country's elite is firmly focused on large-scale foreign investment ventures as it pursues a strategy of increasing inbound tourist numbers regardless of sustainability and conservation criteria. In 2016, for example, the Cambodian Ministry of Tourism listed only 56 'official' CBT ventures (Chan, 2016). The Asian Development Bank-funded Greater Mekong Sub-region Tourism Development Program 2005–2008 included CBT as one of three main areas for action in Cambodia, and it financed the Tourism Ministry to employ an international expert on community-based ecotourism (CBET) with a team of local counterparts. Despite vigorous pursuit of a target of 100 projects and numerous project proposals, however, less than 40 were initiated and within five years less than ten were still extant. Given the poor history of CBT implementation, the Asian Development Bank (ADB)-funded Greater Mekong Subregion (GMS) Tourism Infrastructure for Inclusive Growth Project, 2014–2018, the successor to the GMS Tourism Development Program relinquished rural CBT and focused on infrastructure, constructing a marine ferry terminal and other facilities to boost tourism in southern Cambodia, including upgrading a limited number of rural roads and wastewater management in areas with well-preserved Khmer and colonial architecture, including Kampot province and seaside towns in Kep and Koh Kong provinces. While couched in the language of poverty alleviation and rural development, these projects tend to provide major benefits for the 'big end of town' and little for responsible rural CBT development. The 2016–2022 Cambodian Tourism Strategic Directions Plan (Table 11.1) focused very heavily on numerical targets but neither rural tourism and responsible tourism nor CBT/CBET were specifically highlighted. Its reference to 'enhanced product quality' related to more four- and five-star hotels than alternative responsible tourism development.

Table 11.1 Strategic directions of Cambodian government's tourism sector

Strategy	Strategic directions	Targets
Cambodia	1. Enhance tourism product development and product quality	1) 7.5 million international visitor arrivals (IVA)
Tourism	2. Improve marketing and promotion	
Development	3. Improve travel facilitation, transportation,	
Strategic	and regional and international connectivity	2) USD5 billion international
Plan 2012–2022	4. Improve tourism safety and management of negative environmental, social, cultural, and economic impacts	visitor expenditure
	5. Strengthen legal systems and management mechanisms	3) 1 million new jobs
	6. Human resource development	

Source: Schipani (2017).

Under these circumstances, a number of I/NGOs, such as WWF, the Wildlife Alliance, the Cambodian Community Based Ecotourism Network (CCBEN) and MB have taken up the challenge to fill the relative void.

'MlupBaitong'

MlupBaitong ('Green Shadow' in English), was founded in 1998 by a British NGO to confront the challenge of deforestation in Cambodia, which between 1970–1992 saw forestry cover decrease from 70% of the country to only 40% (Dennis & Woodsworth, 1992). As an advocate of environmental conservation, its objectives emphasized educating the general public on sustainability and protection of natural resources with a strong focus on forestry (Va et al., 2012). In 2001, MB localized, became independent, established its own board of national directors and drew up a set of by-laws which moved to emphasize community benefits equally with environmental conservation and forestry protection.

One of its first projects examined deforestation in Kampong Speu Province located about 100 km West of Phnom Penh, bordering the Cardamom Ranges and including the Kirirom National Park within its jurisdiction. Once heavily forested, parts of the Cardamoms and MB targeted Chambok Commune, located on the eastern slope of the Kirirom highland, had been heavily depleted. The Commune administers nine separate villages consisting of more than 540 households with a total population of 3840 (2016). A survey carried out by MB in 2001 revealed that 94% of the households in the nine villages were engaged in various forms of extractive industries. They included logging for timber and firewood, charcoal production, non-timber forest products collection and hunting. Bamboo shoot/pole harvesting was one of the few sustainable activities because of its annual growth cycle after cutting. There were 72 charcoal kilns within the project site and they consumed several hundred trees each day. The soil of the area was relatively infertile; the villagers could produce only one crop of rice per year, and in the struggle to survive they had reduced the timber resources outside the Park so that illegal logging and poaching inside the Park's boundaries were an increasing problem (MlupBaitong, 2003).

Rather than pursue single-mindedly environmental controls and regulations, MB turned to the potential for ecotourism to engage the communities in more responsible livelihoods. To empower Chambok community members to actively participate in the sustainable management of their natural resources, to reduce poverty, and improve livelihoods, three main objectives were articulated:

(a) protection of forests and natural resources;
(b) provision of income-generating alternatives for poverty-stricken and forest product-dependent communities through ecotourism; and
(c) education of local people and visitors about environmental conservation (Va *et al.*, 2012).

In pinpointing ecotourism as an alternative, MB brought in a Japanese expert to assist them for one year since there was no in-house expertise. As attractions, a 30-m high waterfall, a cave with three species of bats, the beauty of remnant forested mountains, valleys and streams for nature treks, and the village way of life (culture) were identified (see Figure 11.1) and MB carried out a preliminary feasibility study which suggested that CBET was a viable alternative.

In 2002, MB took their study to a meeting of the Chambok Commune Development Council where they first proposed the idea of ecotourism. This was a novel idea for the communities so MB ran three workshops to explain the need to protect the forest resources and what ecotourism was, and gained their support. With MB's guidance, the communities elected a 13-member Ecotourism Management Committee (EMC) representing each of the nine villages, which included a minimum of three women as mandatory, two advisers from the Development Council and Kirirom Park management. One of their first decisions was to enact an immediate ban on logging and charcoal production to prevent further deforestation.

At this point, MB embarked on an innovative approach to protect the area and engaged key government stakeholders including the Ministry of the Environment, the administration of Kirirom National Park and the Province Governor to establish an integrated 'Protected ecotourism forest site'. A series of negotiations resulted in the Ministry of the Environment and the Provincial Government agreeing to a complex lease agreement that gave the Chambok EMC responsibility for 161 hectares within Kirirom National Park, 800 hectares bordering the Park as a 'Forested community protected area', and a further 300 hectares designated as a 'Community Forest' (a common usage reserve where some non-timber forest products (NTFP) collection and harvesting would be permitted). In

Figure 11.1 Natural attractions of the Chambok area: Kirirom Waterfall, rapids, flowers of the rainforest and the trail to the waterfall

2005, a further 131 hectares were added to the community protected forest for a total of 1,391 hectares, a contiguous area of almost 14 sq km (Chambok Ecotourism Site, 2017) The novelty of this arrangement is difficult to over-state since the entire project could not have proceeded without Chambok having approved legal access to the forests for its ecotourism venture. The lease was the first of its kind in Cambodia and has subsequently been used as a model for similar projects in Thailand and Laos (Men Prachvuthy, 2006; Va *et al.*, 2012).

The Chambok Community-Based Ecotourism Venture

The Chambok ecotourism project was formally inaugurated on 4 January 2003, in a ceremony attended by the Governor of the Province and senior district, provincial and ministry officials, including a representative of the Royal Government of Cambodia's Ministry of Tourism from Phnom Penh (see Figure 11.2). Before the project could be established, however, the area had to be assessed by the Cambodia Mine Action Committee (CMAC). Once declared safe, MB organized the development of the site into five main components. The first was infrastructure and maintenance, which included initially 7 km of nature trails, (since expanded to more than 15 km), small lengths of boardwalks and stairs, a

Figure 11.2 Welcome to Chambok – The Cambodian Minister for Tourism, Dr Thong Khon, is presented with a Cambodian sugar palm, the national tree of Cambodia, by the Executive Director of MlupBaitong, Mr Va Moeurn and a guide from the Chambok Ecotourism Management Committee, Chambok, July 2013

number of small bridges, a track into the forest for bullock cart rides, toilets and rest shelters at strategic locations along the trails, an entrance booth, a reception-and-information centre for arrivals and a small tree nursery to regenerate degraded areas of forest and a car park for visitors. Labour was provided by the villagers, with MB providing small wages and 'food for work' based on nutritional standards set out by the UN World Food Program (Va *et al.*, 2012).

The second component was capacity building, and MB ran training workshops on guiding, maintenance, environmental knowledge, project planning, management and book-keeping. MB also sent members of the EMC to Thailand to increase their understanding of how to implement ecotourism ventures. The third component focused on marketing and setting up income-generating services such as a system for ticketing, car parking, food preparation, ox-cart rides, guided tours, souvenir stalls, cultural performances and homestays.

The fourth component was designed to maintain the environmental health of the site, from daily patrols of the forests to prevent illegal logging and hunting, and fires to replanting trees, solid waste management with installation of rubbish bins and a safe disposal system, and control of foraging domestic stock such as goats, pigs and cattle. The daily patrols are carried out on a roster basis by villagers themselves.

The fifth component established Women's Self-Help Groups (WSHGs) for micro-ventures such as bicycle hire, homestays, souvenir vending, and food preparation (meals and snacks) for visitors. Because of the complete absence of any restaurant within the area, by 2005 the demand for this latter service resulted in the opening of the 'Women's Forest Café' with a capacity to provide lunches for up to 100 visitors. Homestay training was provided, but uptake was very slow and by 2006 only four households had upgraded their facilities (bedroom, bathroom) to open their doors.

Early difficulties

In common with many CBT projects in many countries, Chambok struggled in its formative years to meet expectations. It was unable to achieve sufficient tourist visitation to offset the complete ban on logging and charcoal production, despite hosting 11,000 tourists per year in both 2003 and 2004. A study carried out in 2005 on the 127 households in Chambok and Boeng, the two villages most directly involved in ecotourism at the time, revealed that the per household income on average dropped from between USD200–USD500 p.a. for 94% of families previously engaged in forestry extraction (especially charcoal) to just USD26 p.a. (Men Prachvuthy, 2006). More than 95% of the visitors were Cambodian, almost two-thirds schoolchildren on educational visits (a major part of MB's outreach programme on forestry conservation), and per capita spending of visitors was very low. Rather than ecotourism alleviating poverty, the bans ran the risk of creating new poverty. In hindsight, it can be seen that there should have been a transition period where logging and charcoal production were phased out over several years, and a plantation forest established to allow sustainable logging and charcoal production in future years.

The project survived its early years not because of direct tourism-generated income, but mainly because MB, with funds *inter alia* from the EC and United Nations Development Programme (UNDP), was able to provide wages for all those engaged in various services and activities, 100 men and 200 women on a roster basis, and to finance various projects such as trails, construction of the visitor information centre (at 92 sq.metre substantial building), the women's café, toilets, the tree nursery, and so forth. MB continued underpinning the project until 2009 when it reduced its involvement to an advisory capacity, and left full management in the hands of the local committee on the grounds that the ecotourism venture had become self-supporting. By this time, MB's investment totalled USD226,000 (Va *et al.*, 2012). Foreign donors included UNDP, EC, ADB, Oxfam, Scottish Catholic International Aid Fund, the Keidanren Nature Conservation Fund, the Blacksmith Institute, the Canada Fund and the McKnight Foundation (Va *et al.*, 2012). Neither the Cambodian Government nor the private sector contributed financial support.

By 2009, it had become clear that while ecotourism was providing a reasonable income for perhaps 20 families, and was contributing several thousand dollars each year to the Community Fund so that all benefited from the Fund's activities (such as a small reservoir drawn from the waterfall river to provide the communities with year-round piped water), on average, tourism generated a supplementary contribution of only 40% of household income, and was not sufficient by itself to maintain family living standards above poverty. In order to make ends meet, there was an inevitable resumption of some illegal logging for timber and firewood although charcoal production remained inactive, and the logging took place outside the designated 'ecotourism community forest'.

Sustainability of the Chambok ecotourism venture

As part of an integrated approach to forest conservation, in association with ecotourism, MB provided for improved incomes from the sale of non-timber forestry products, which were (and continue to be) sustainably harvested and conserved, based on community-approved controls. Farming incomes also increased because of tourist consumption through the Women's Forest Café with guidance from MB in better farming techniques and new crops. MB also provided sound training for the EMC to apply for project funds from the pool of aid donors mentioned above, and the Committee's success in maintaining a steady flow of such funds have contributed to continuing support over the years by the community at large for Chambok's ecotourism venture. New projects (e.g. improved access road to the site, expansion and upgrading of trails, a major upgrade of the visitor information centre and the WSHG Forest café, and the permanent community water supply) have continued to provide employment opportunities for many community members, in addition to direct income generated by tourists (see Figures 11.3 and 11.4).

Currently, Chambok has 40 households engaged in homestays and it continues with its various activities including the WSHGs and forest café, tree replanting, guided rainforest tours, oxcart rides, bicycle hire, and so forth (see Figure 11.5). Its visitation has plateaued at around 12,000 p.a. but by working with several commercial tour operators in Phnom Penh, about 20% of its visitors are now foreign tourists who stay for an average of two nights with a daily spend estimated at USD30, against locals who are almost entirely day visitors with an average daily spend of less than USD10 (MlupBaitong, 2018) The total benefits arising from MB's intervention in community development, ecotourism and NTFPs have resulted

Figure 11.3 Chambok CBT infrastructure – expanded and refurbished visitor information centre, 2013. Extended Women's Group Rainforest café

Figure 11.4 The mini-reservoir funded by the European Community

Figure 11.5 Chambok ecotourism commercial activities – Chambok homestay traditional village house, Chambok ox cart for hire

in stability for the majority of households and, together with other income streams, has lifted them above the poverty line of USD1.90 per capita per day (MlupBaitong, 2017).

The timeline shown in Figure 11.6 enumerates the major milestones in MBs successful work with Chambok from the beginning to 2017, including recognition of its responsible rural development mantra based on nature conservation with a number of awards. These include the Cambodian Civil Society Partnership award for 'Promoting community environment issues and ecotourism' 2006; a Bronze Medal by Cambodia's Prime Minister for achievements in Community-Based Forestry and Protected Area project of Chambok; US Aid Cambodia's 'Hidden

172 Theme 3: Tourism That Takes Place in Rural Areas

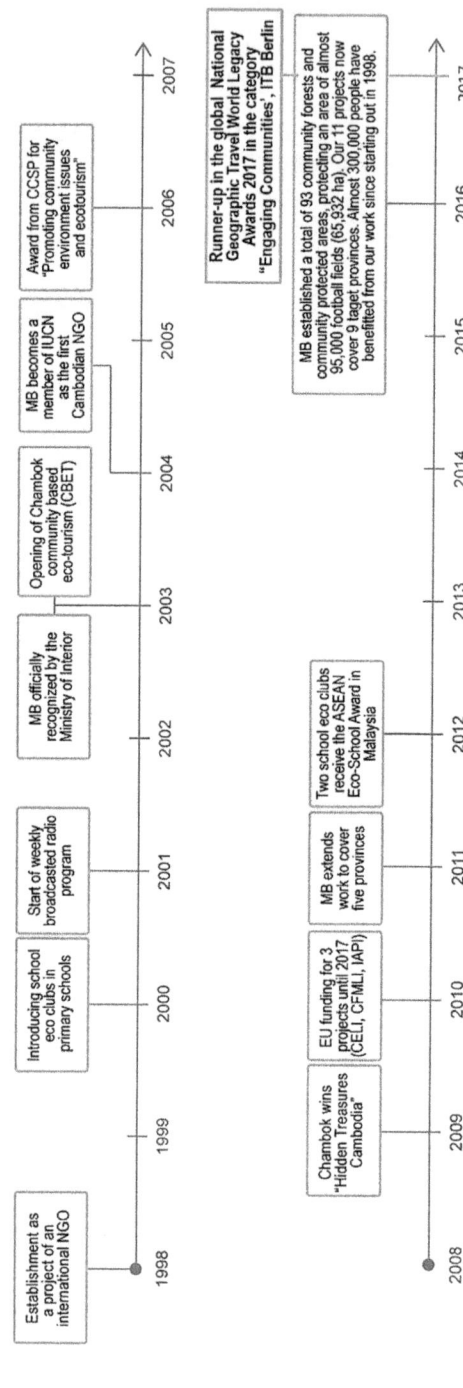

Figure 11.6 Genesis of 'MlupBaitong', 1998–2018
Source: MlupBaitong (2017)

Treasures' Micro, Small and Medium Enterprise Project award 2009; The 'TO DO! 2013 Award for Socially Responsible Tourism', International Tourism Boerse (ITB), Berlin, Germany for Chambok Community-Based Ecotourism; the ASEAN Homestay Award for Chambok 2016; and the runner-up for the National Geographic Travel World Legacy Awards 2017 in the category 'Engaging Communities'.

Conclusions: Chambok as Responsible, Rural, Community-Based Ecotourism

In applying the definition of responsible tourism used for this book, i.e. 'Responsible tourism is tourism which minimizes negative social, economic and environmental impacts, generates greater economic benefits for local people and enhances the well-being of host communities' (Goodwin, 2012), it can be argued that Chambok meets all of these requirements. It is located more than 100 km from the capital of Cambodia, more than 30 km along a secondary road off a major highway and consists of nine rural farm-centred villages. The model of ecotourism introduced by MlupBaiting has minimized environmental impacts, and has enhanced social well-being, although in the beginning, it created some negative economic impacts with its immediate ban on logging and charcoal production without planning a transitional phase and establishing forest plantations. Without the immediate ban, however, environmental degradation would have continued on a much larger scale until the detrimental activities were completely phased out. Further, the tree planting scheme and sustainable management of the harvesting of NTFPs have countered the initial reduction in incomes and improved the overall state of the natural environment of the Chambok area. In terms of social impacts, the Capetown definition of responsible tourism does not specifically mention gender equity, often identified as a key factor in sustainability (Scheyvens, 2000), but one area of outstanding success in Chambok has been MB's introduction of the WSHGs and micro-financing. The aspect of respect and conservation of intangible cultural heritage has also been addressed in the activities that are presented by the Chambok community to visitors, including the transfer of traditional knowledge (TK) by guides to visitors trekking along the rainforest trails. In addition, it has introduced democratic processes into the management of the Chambok ecotourism venture through the elected EMC and the way in which households are rostered on a revolving basis to undertake activities and services for visitors and thus benefit directly from the tourist dollar. This is another element on which the Capetown Declaration is also silent, but again in terms of responsible tourism, democratic processes are viewed as a key component of sustainability and social wellbeing (Honey, 2008). In this context, Walter (2011, cited in Reimer & Walter, 2013: 123), suggests that ecotourism needs to move beyond consideration of economic and environmental impacts to include:

> (a) principles of local participation, control or ownership of ecotourism initiatives; (b) a focus on environmental conservation and local livelihood benefits; (c) the promotion of customary and indigenous cultures; and to some extent, and (d) the promotion of local and indigenous human rights and sovereignty over traditional territories and resources.

All of these principles are embedded in what MlupBaitong (2016: 5) calls its 'Complementary Approaches':

> 'The **rights-based approach** is used to build human rights awareness among target groups to enable them to advocate for their rights.
>
> The **empowerment approach** builds people's capacity and competence as individuals and participating members of groups and communities to achieve results for themselves.
>
> This includes **assessing their own needs and rights**, developing a vision for change, and planning, implementing, monitoring and evaluating their projects.
>
> To ensure that women participate in and benefit equally from all development activities, we use the **gender-based approach** in all we do.
>
> In many of our projects, we also apply the **micro-project approach** to support income generating and environment protection initiatives.
>
> Finally, the **integrated approach** interlinks aspects and components of individual projects such as empowerment, community-based forestry management, community-based ecotourism and livelihood improvement, to attain synergy effects and allow tailored initiatives for each community'.

It should be noted that MB's activities extend well beyond ecotourism to incorporate more than 150 other projects covering forestry, primary and secondary school outreach education programmes, community farming and other many micro and small community-based ventures. Following the success of Chambok, MB has established four other community-based environmental conservation and ecotourism sites in other parts of Cambodia that are progressing well. In short, the integrated approach adopted by MB, which has combined commercial activities through ecotourism with forest protection and conservation for community wellbeing, has provided a model for responsible rural tourism in Cambodia that neither the public nor private sector has emulated and which in part exceeds the requirements of the Cape Town Declaration on responsible tourism. Additionally the game-changing 'Forested community protected areas' and 'Community Forests' have managed to reconnect the local communities with their surrounding natural environment to re-establish a harmonious relationship between people and nature. In 2018, such areas have been formally recognized as Other Effective Area-Based Conservation Measures (OECMs) (IUCN WCPA, 2019). In essence OECM resonate well with areas that are revered and conserved by the local community such as sacred natural sites, community forests and other areas with strong cultural and spiritual values. Although large parts of formally protected areas have suffered encroachment, the community protected areas and forests that support sustainable livelihoods have provided a new approach in developing responsible CBT using the MB model that is being replicated not only within the country but also in Thailand and Laos.

Author's note. Dr Sofield was Team Leader for the Asian Development Bank-funded 'Greater Mekong Subregion Tourism Development Programme', based in Cambodia from 2005–2008. He participated in ADB assistance to Chambokin 2006 and contracted MlupBaitong to implement another ADB-financed community based ecotourism project in SteungTreng Province, Cambodia, in 2007. All photographs were taken by the author.

References

Chambok Ecotourism Site (2017) Chambok Ecotourism Site: Discover nature and rural Cambodian life. See https://chambok.org/ (accessed 9 June 2017).

Chan Socheat (2016) Sustainable development and tourism in Asia: a case from Cambodia., *International Workshop on Sustainable Development and Tourism*. Center for Sustainable Development Studies, Toyo University, Hakusuna, Tokyo, 27 February 2016. Phnom Penh: Ministry of Tourism.

Dennis, J.V. and Woodsworth, G. (1992) *Environmental Priorities and Strategies for Strengthening Capacity for Sustainable Development in Cambodia*. Phnom Penh: UNDP.

Goodwin, H. (2012) Ten years of responsible tourism: An assessment. In *Progress in Responsible Tourism*. Oxford: Goodfellow Publishers.

Honey, M. (2008) *Ecotourism and Sustainable Development: Who Owns Paradise?* Washington: Island Press.

IUCN WCPA (2019) Guidelines for recognising and reporting other effective Area-based conservation measures. Switzerland: IUCN Publication.

Men, P. (2006) Tourism, poverty, and income distribution: Chambok Community-based ecotourism development, Kirirom National Park, Kompong Speu Province, Cambodia. *Journal of Greater Mekong Subregion Development Studies* 3 (1), 25–40.

MlupBaitong (2003) Report on natural resource utilization and food security. Phnom Penh: MlupBaitong.

MlupBaitong (2018) Annual Report, 2016. See https://www.google.com.hk/search?q=Mlup+Baitong+(2016).+Annual+Report%2C+2016.&oq=Mlup+Baitong+(2016).+Annual+Report%2C+2016.&aqs=chrome..69i57.6232j0j7&sourceid=chrome&ie=UTF-8 (accepted 20 June 2018).

MlupBaitong (2018) Our history and awards. See http://mlup-baitong.org/our-history/ (accessed 20 June 2018).

Reimer, J.K. and Walter, P. (2013) How do you know it when you see it? Community-based ecotourism in the Cardamom Mountains of south-western Cambodia. *Journal of Tourism Management* 34 (1), 122–132.

Scheyvens, R. (2000) Promoting women's empowerment through involvement in ecotourism: Experiences from the third world. *Journal of Sustainable Tourism* 8 (3), 232–249.

Schipani, S. (2017) *Tourism Sector Assessment, Strategy, and Road Map for Cambodia, Lao People's Democratic Republic, Myanmar, and Viet Nam (2016–2018)*. Manila: Asian Development Bank.

Sofield, T.H.B. and McTaggart, R. (2005) Tourism as a tool for sustainable development in transition economies. *DFID/GRM Conference on Development Learning in Transition Environments*, London, 18 October 2005.

Sofield, T.H.B. (2009) Year Zero! From annihilation to renaissance: Domestic tourism in Cambodia. In S. Singh (ed.) *Before Tourism: Explorations in Holidaymaking and Journeying in Asia* (Ch 7, pp. 151–180). London: Earthscan Publishing Ltd.

Transparency International (2016) Corruption Perceptions Index 2016. See https://www.transparency.org/whatwedo/publication/corruption_perceptions_index_2016 (accessed 11 July 2018).

United States Government (2018) *The World Fact Book: Cambodia*. See https://www.cia.gov/library/publications/the-world-factbook/geos/cb.html (accessed 14 July 2018).

Va, Moeurn, Lay, Khim and Chhum, Sovanny (2012) Good practice in the Chambok community-based ecotourism project in Cambodia. See https://ysrinfo.files.wordpress.com/2012/06/chombok1.pdf (accessed 16 May 2018).

Walter, P. (2011) Gender analysis in community-based ecotourism. *Tourism Recreation Research* 36 (2), 159–168.

World Bank Governance Indicators (2017) The Worldwide Governance Indicators (WGI) project. See http://info.worldbank.org/governance/WGI/#home (accessed 11 July 2018).

World Bank Group (2016) *Cambodia Economic Update: Improving macroeconomic and financial resilience*. Phnom Penh, Cambodia: World Bank Publication.

12 Can Tourism be a Success Story? Stakeholders' Management and Narratives of Rural Tourism. Reflexive Analysis of a Tourism Project in Timor Leste

Frederic Bouchon

Introduction

Tourism is considered a tool for socioeconomic development and cultural preservation by governments and private agencies. But the tensions between promises of sustainable development made to a community and the actual results and their sociocultural impacts have been an ongoing challenge (Cabasset-Semedo, 2009; Scheyvens & Russell, 2012; Tolkach, 2013). This is especially acute in the case of smaller, less developed and peripheral countries in which problems of tourism development stem from inadequate power relationships between stakeholders. While many studies investigate the perspectives of tourists, few examine host perspectives. The networks of influence on which rural tourism is built remain largely overlooked with a scale of relationships that involve global, national and local stakeholders. This situation is compounded in the case of rural heritage tourism with few studies analysing the relationship between tourism, stakeholders and what makes the rural space a destination. This extends to the understanding of the imaginaries on which heritage tourism relies and how it reshapes the community's sense of place.

This study is based on a field research in Timor-Leste, one of the world's most recently established nations and least developed small island states. Government and non-governmental organizations (NGOs) are trying to promote tourism in order to offer jobs, create business and income for the local economy, and to improve regional imbalance (Government of Timor-Leste, 2011). Moscardo (2011) emphasizes the importance of capacity building for sustainable outcomes of tourism in the case of developing countries. The study was conducted in conjunction with a tourism-capacity building project in a boutique hotel created in a renovated heritage site. This project provided a suitable platform to embark on a study of the narratives of the place, and their importance in the reshaping of the community through the tourism lens. The study of 'narratives of tourism' constructed by location stakeholders allows reflection on the networks in place and how a tourism development project reflects on 'glocal' issues in a Timor Leste village. It uses an ethnographic approach together with a reflexive analysis from the author. The objectives of the study include the review of the capacity-building programme, and how a human agency makes arrangements with the changes brought by tourism, expected and real. It reflects on the narratives of tourism from multiple angles: heritage, postcolonial and the governance dissonances. This study places emphasis on sociocultural issues related to the portrayal of heritage and its tourism representations. The first part reviews the literature on tourism in rural communities in developing countries. It also reviews the narratives of tourism. It then analyses the particular project in Timor Leste, and discusses the current outcomes.

Background Information

Narratives of tourism, 'glocal' imaginaries

'Tourism narratives' is an expression of polysemic nature. It can refer to the discourse on tourism as the grand narrative of 'tourism'; it can be tourist stories; it can be host stories; and it can also be tour guides performing narratives. The control of the story is key; who tells the story to whom is an important question (Harrison & Hitchcock, 2005). The memory and the stories that a place evokes is the key to its reality from the user's perspective – in our case, the stakeholders. Salazar and Graburn (2014: 17) refers to the imagery of the place, contested narratives and the power relations implied:

> The failure (..) to understand how imaginaries are embedded within local, national, regional, and global institutions of power restricts their ability to determine the underlying forces that restrict some tourism practices, and not others, some imaginings, and not others, and that make possible new hegemonies in new fields of power.

In addition to this, technological changes have put flows of people, ideas and capital at the centre of a network society challenging territorial hierarchical systems (Castells, 2000; Sassen, 2007). Some argue that tourism creates difference, aggressively re-imaging, re-constituting and appropriating heritage or culture of the place, insisting on symbols of being uniquely local in a futile effort to downplay the global serialization of cultural consumption (Bouchon & Lew, 2014; Hannigan, 1998; Martínez Gimeno, 2007). Franklin (2004) says, 'It was not difference and the extraordinary that created tourism but the opposite, the extension of belonging, the prospect of taking up a place in the new national cultures that beckoned them'. The attention given to narratives contributes to unearthing multiple realities. Ontologies of tourism are evolving with glocal place narratives being put together by tourism actors and networks. Franklin (2004) emphasizes the 'remaking of the world anew as a touristic world'. Cultural links also create narratives with proximity regardless of space and distance as in the case of Timor Leste: due to their shared colonial history, Timor Lester shares more similarities with Brazil than with its closer neighbours

Rural tourism in small developing countries

Rural tourism in developing countries is a complex phenomenon, ranging from ecotourism, agrotourism, community-based tourism (CBT), to cultural tourism (Tane & Thierheimer, 2009), while large resorts, developed in rural areas are excluded, being urban enclaves. 'Rural tourism' is seen as a tool generating moderate economic growth, maintaining the sociocultural fabric, while allowing locals to venture, with limited capital and training, into new economic activities (Kalsom, 2008; Wilson et al., 2001). The issue of power, between centre and periphery shows political and social differences. Central governments view tourism as a solution to high unemployment, isolation and the lack of industrialization in rural communities. External projects such as resorts, or museums, originated from an international conservation agency, a donor, or a private tourism operator, given the economic impact, are more likely to receive support from the authorities (Scheyvens & Russell, 2012). Cohen and Cohen (2012) mention that, 'tourism typically benefits middle classes, as opposed to the poor, with the latter most affected by the negative impacts of tourism'. That is particularly important for rural areas, in that the entrepreneurs are more likely to come from the city, equipped with understanding of tourism and business and have the necessary capital. In a contrasting view with the government, residents, businesses and NGOs see 'greater involvement and control from local populations on tourism, and a strong link to agriculture'(Cabasset-Semedo, 2009).

Yet communities are likely to have limited financial, business, human resources and limited management systems to initiate tourism development. Rural tourism as a process is nurtured through the cooperation and collaboration among associated stakeholders such as government, NGOs, communities, tourists, volunteers and private sector through a strategic approach, such as training. The approach to tourism development in rural areas is often mixed, in terms of ownership and origins of the tourism development initiatives. Small-scale private endeavours do not necessarily qualify as CBT, and a larger framework of local rural tourism management could help in integrating the local situation. Likewise, for the communities, adaptability of this exercise through mutual cooperation and partnership equally plays an important role for the successful outcomes. Even after these elements are addressed, the questions of control, authority, agency and response to it determine the legacy of projects. A number of authors (Sakata & Prideaux, 2013; Salazar, 2010; Scheyvens & Russell, 2012) argue that successful rural tourism developments are always linked with inequitable power relations between stakeholders (locals and outsiders). The negotiation that takes place between the needs of service quality and the expectations of the local community and the vision of the future would be at the centre of a local community development master plan. That requires a redefinition of stakeholders and their roles.

Managing stakeholders' challenges

Several authors have placed the emphasis on political contexts, local leadership and the willingness of the community members to participate in discussing the success of local development through tourism (Sakata & Prideaux, 2013; Salazar, 2010). The managerial and technical support provided by NGOs, government, institutions or donor agencies, and the sufficient skills developed to implement appropriate exit strategies are considered as a backbone for the foundation of every rural tourism project. Several authors also underline the importance of stakeholder management for any rural community development through tourism (Currie & Turner, 2011; Yang & Pandey, 2011). Understanding why a project succeeded or failed lies primarily in an evaluation of attitudes to the project within the local community. It remains unclear who the stakeholders are in a tourism project. Some take the viewpoint of local residents.

However, the definition of stakeholders is complex. Some residents have migrated to other regions or countries and come back as investors, or still claim a legitimacy in decisions pertaining to the future of the community. Remote rural and peripheral communities have to deal with the national, political, social and economic structures that may limit the participation of local people while being involved in tourism projects. Simpson (2008) highlights that often NGOs and development agencies who are supporting CBT ventures have limited knowledge about the needs of the tourism industry. King and Tolkach (2015) stress that the power 'lies with the tourism sector and not with communities'. In other words, CBT operators have no choice and therefore comply with norms and expectations that may sustain neocolonial aspects. That raises issues of defining stakeholders and their legitimacy. The host community is often perceived as a 'romantic and essentialist' notion (Sin & Minca,

2014). The fragmentation of the host community is emphasized in the Social Exchange Theory, in which residents with no ties to tourism (e.g., farmers or traders, craftsmen) are not supportive of tourism activities but have a positive perception of tourism development if it leads to personal financial gain (Lee, 2013; Nunkoo & Ramkissoon, 2010). Some locals, after living overseas, play the role of change advocates. Their legitimacy might be challenged, and they need the help of the authorities to gain support from the community and champion the idea.

Timor Leste, tourism destination and nation building

Timor-Leste is a young developing island country ranked 133rd out of 188 countries by its Human Development Index (UNDP, 2015). The society is largely agrarian with 70% of the population living in rural areas. Agriculture (coffee and rice) generates 27% of the gross domestic product (GDP) while oil and natural gas are the main economic contributors. Timor-Leste still depends on foreign aid. Since independence in 2002, tourism has been seen as a solution to diversify the economy and generate employment. However, the number of tourists remains low. In 2015, overall there were only 65,000 arrivals (ADB, 2015). The majority came for work and business purposes while only a minority of travellers visited primarily for holiday purposes (17%). Actually it relies on 'what resembles a kind of tourism for locals' (Cabasset-Semedo, 2009), with a large number of foreigners having an extended stay and working for NGOs and aid agencies. There has been growth in ecolodges, homestays in the countryside that attract small numbers of ecotourists and sports tourism. Nevertheless, Timor-Leste remains a relatively uncompetitive and expensive destination compared with Indonesia. It is perceived as unknown, with limited access, sites, little tourism infrastructure and poor branding. Currie and Turner (2011) observed the contesting views between government and communities on tourism development in Timor Leste. 'Both the stakeholders and the government want tourism, but (…) the government wants quick development of mass tourism, the local communities and businesses want a cautious targeted programme of slow development'. The development of any tourism is desired as a source of national revenue, overshadowing the diverging interests of marginalized peripheral communities, who advocate 'rural tourism'.

Study Approach

This study originated with my participation in the delivery of a tourism-capacity training to the employees of the first international hotel in a refurbished Portuguese fort, in the West part of Timor Leste. Although small, this place holds particular national historical significance and has attracted the attention of several donors over the years. The government of Timor-Leste (2011) has included the fort restoration project in its National Strategic Development Plan 2011–2030. The methodology follows an ethnographic approach adopting a reflexive account to understanding the 'glocal' dynamics of tourism and its embedment in a remote community. A 'glocal ethnography' as defined by Salazar (2010: 191) is used as it allows a study of tourism development, imaginaries from the multi-scalar views of stakeholders, while using the capacity-building training as the anchor to the project. Data collection originated from site observation and informal and formal conversations with the participants, community members and the outside stakeholders involved in the project. This allowed an understanding of how stakeholders negotiate the promises of tourism development.

A thematic analysis was made, based on the narrative of the project participants and my reflexive practice. May and Perry (2011) argue that 'a reflexive practice can be framed as standing guard against those who conflate the model of reality with the reality of the model and so allow clarification in the endogenous realm of scientific practice to unproblematically spill over into the referential realm'.

The boundaries created by models brought to places in which the problems have not been diagnosed are essentially due to the lack of reflexivity on the area studied. I used the notes taken during the course, various observations and reflective notes to dwell on the issues of narratives of tourism and hierarchies. I was involved as a trainer and the need to use reflexivity was felt, as my position and role evolved during the six months of training. Being involved in the role of a researcher but also being part of the delivery project allowed me to reflect on my position in this study, as an academic from France who has been living in Asia for more than 15 years. This has progressively given me an intercultural perspective to understand different positions and worldviews. Belsky (2004) argues that 'tourism researchers rarely speak directly about the values that influence their choice of topics and the research methods they employ'.

The politics of tourism and the reality of power in every layer of tourism are muted because tourism research has not been particularly rich with ethnographic detail on the particular people and place in which the politics are embedded in the assessment of the capacity-building programme. Researchers' perspectives also draw on the perception of the challenges faced by rural communities. The study uses verbal and nonverbal snapshots that are the basis of the tourism narratives and the interpretivist analysis of the multiple scales of encounters generated by the irruption of a heritage boutique hotel in a rural farming community. Pereiro (2010: 175) indicates that fieldwork produces knowledge based on the researcher's experience in a reflexive account. Riessman (2007) underlines the importance of the role played by the researcher, the derived narratives and language. In other words, recognizing the subjectivity of one's frame of thought and his socially constructed realities enables him to adopt an honest and perhaps humble positioning vis-à-vis his study. This moves away from the essentialist approach of subject–object while experiencing intercultural contact with the purpose of knowing 'otherness' against the possibility of multiple interpretations of reality (Hall, 2004; Mura, 2016). In line with the call of Botterill (2007) to include the 'situated voice' in tourism, I will use the context of this project to call for more reflexivity.

The Tourism Capacity-Building Programme

Context of Balibo community

Tourism in Timor Leste is still at an infancy stage. In Balibo, tourism remains limited, although the location of the small town and its history make it a natural destination. The small town with a population of less than 3000 people lies 400 m above sea level and dominates the western coast. The strategic location controlling the whole Bobonaro district hinterland made it the site for a Portuguese garrison and fortress in the 18th century. The fortress infrastructure has remained intact. In 1975, a few weeks after the Portuguese colonists decided to let Timor takes care of its own future, undercover invading Indonesian troops entered the country through Balibo. The story of five Australian journalists murdered while covering the event had a deep impact overseas and the case of the 'Balibo-5' channelled sympathy to the struggle of Timorese towards independence for 24 years. The history of Timor Leste summarized by the struggle towards independence continues to capture people's imagination (Cabasset-Semedo & Durand, 2009). Tourism is used as a vector of nation building, in any country, but especially in a young post-colonial country (Ren *et al.*, 2010).

The Timorese culture was reshaped by Suharto's government attempt at subjugating its culture under assimilationist ideals, while simultaneously celebrating an idealized pacified Indonesian Timor. Timor-Leste is now trying to develop itself as a niche cultural tourism destination since its independence from Indonesia in 2002. In an effort to move away from the tormented past, the Balibo House Trust (BHT) was established in October 2002 by the Victorian State Government and the families of the Balibo-5. BHT decided to expand its assistance to the Balibo community and was awarded a 30-year lease for the historic Balibo Fort. It supported the renovation of the fort building, and raised funding for the layout of a heritage boutique hotel.

Training tourism professionals in a rural area: the capacity-building programme

The lack of assistance from the government sector and the numerous hurdles facing the private sector have confined tourism to an infancy stage, with limited infrastructure beyond the municipal limits of Dili, the capital city. In this light, developing a private tourism infrastructure outside of Dili could be viewed as a daring venture. This programme aims to lay the base for tourism infrastructure in the Bobonaro district with a hotel catering for international tourists. Timor Heritage Hotel (THH) is the private company managing the Balibo Fort Hotel (BFH). The aim of the heritage hotel project is also to support the local community and to create a spill-over that enables CBT. The conditions set by the donors in the call for tender sought to avoid the pitfalls generally attributed to capacity-building programmes: trainers-centre learning, contents and delivery inadequate with local needs, and no monitoring of results over time. The boutique hotel comprises eight rooms with en suite, bathrooms each of which has hot water supply, air-conditioning, TV and internet access. Hotel facilities include a restaurant, bar, meeting rooms and gallery, within the walls of the Portuguese fort. The hotel was designed as a boutique hotel, with expectations of international standards. Away from the capital city, the accommodation infrastructure was rudimentary: homestays or Portuguese-era *pousadas*. Modern equipment, quality of the renovation, location and manicured gardens make it on par with the standards of service and expectations in the capital city.

The training programme was designed to develop 20 employees' operational skills and standards in service skills. The programme's participants didn't change much over the six-month period with 18 staff completing the programme, while another three resigned and were replaced. Most of the workforce came from Balibo and the surrounding areas. The majority were hired without prior experience or training in hospitality. The gap of three months between each phase allowed the programme to have feedback in performance based on the objectives, and to understand how the interface between tourists and staff constructed narratives of Balibo as a destination. The training covered all the areas of hotel and restaurant operations, and tour guiding. The capacity-building programme was limited to operational skills development. That objective was achieved, and staff were able to perform their tasks independently upon the completion of the programme. They clearly supported the idea of expanding the tourism industry in Balibo. The programme gave the participants a sense of confidence they did not have before. They took on their roles with great dignity. The ability to reproduce the tasks related to their duty, and follow the standard operation procedures (SOP) was impressive. This was evident from their contact with guests, and the feedback received. Most of the participants had completed basic education, and the final phase became more than just training. A closing ceremony, with a presentation of certificates of completion was organized. The pride taken in being part of this programme and graduating from it was immense and its intensity surprised the team of experts.

From the participants' point of view, this was a moment of pride. 'It's my graduation from an international programme' (J. 29). All participants dressed up in traditional 'tais' costumes. And a group of singers showed other traits of their personality: a strong esprit-de-corps, resilience towards adversity and a quest for peace. Overall, the capacity-building programme was very successful in building the operational skills in hospitality and tourism of employees with no prior training in the service industry. However, some limitations soon appeared. The recruitment of staff, training of staff, and development of SOPs were not done according to any prior experience. The training did not cover business operations, management skills and the recommendation given was to have further training to address these limitations.

Reflections on the Project

The private hotel as an engine for local tourism development

In running a modern hotel defined as a 'boutique heritage hotel' and charging around USD100 per night, the expectations are set: international standards of hospitality. Yet the operations and management of this small hotel of only eight rooms proved complex. Internally, discrepancies were observed at two levels. Firstly, it was between the manager and staff, and secondly between BFH and the head office in Dili (THH). Participants remained throughout the duration of the training. The jobs available in Balibo are essentially agricultural and the hotel offered a serious alternative to the obligation felt by many youths to move to Dili to find a job. A clerical job in the hotel was still considered a better option than no employment at all. The hotel settings were also felt as 'superior, comfortable, nice' (O, 24). The attitude was positive, and the employment in the hotel was seen as prestigious and a source of future opportunities in comparison with other jobs available in the community. This lack of opportunities made the jobs available at the Fort hotel coveted positions. It was interesting to note that these groups of people are the sole beneficiaries of the financial benefits created by tourism through their salary, in a village where most of the employment is in farming and the derived income is irregular. The recruitment of new employees was made in Dili although all come from Balibo. The staff officially started their career at BFH on the first day of training. Most of the employees were relatives, and there was silent resentment of this. 'They choose her (name withheld) because they are same family, cousin' (O, 24). The village politics played an obvious role in the choice, but THH made it clear that the first criterion was based on competencies. The staffing did not change much after the programme completion, and in the following years, illustrating limited employment opportunities in the area.

The manager, although being a Timorese, was not from the district and had lived in Australia. Her status of manager, her Food and Beverages (F&B) experience overseas and her excellent command of English gave her a commending power, which freed her from any bias by community members. Furthermore, the manager stayed in the hotel, and therefore showed commitment to Balibo. In the relations with the head office, the manager faced a gap in communication and sense of direction. The manager had no clear financial objectives, and the procurement planning was obscure. The trade-off for this lack of management system was an asserted authority on the property, based on trust from the head office. There was no evidence of any financial and business objectives beyond the 'breaking even'.

Community development narratives

As in many areas of Timor Leste, NGOs are active in village life. The hotel project was not a fully private, profit-oriented venture, and the NGO made it clear that the scope was a mechanism to trigger and emulate more initiatives from the community. The Fort Hotel management and the Australian-funded NGO that initiated the capacity-building programme emphasized the need to respect the community and some other aspects of responsible tourism. For instance, the project of a swimming pool, inside the fortress was deliberately shunned. The contrast of having a pool in a community facing chronic water shortage was felt inappropriate. Furthermore, the fort was to remain a private operation but a public heritage space. This was clearly expressed with an 'open door' for the community to visit the site. Most importantly, the employment provided to community members was the strongest point to gain respect and trust from the local community. Nonetheless, the question on how to sustain the business for a longer duration was a challenge due to the lack of understanding about tourism management. Hence, continuous discourse was critical to educate the community on responsible tourism.

The responsible tourism discourse by the NGO was received positively especially since much education and training work had been done in the community and had generated expectations. Locals were expecting jobs from tourism. The limits to the project were in expanding training beyond the hotel employees, and in ensuring skills training in trades for maintenance, business, management and communication. The notion of tourism management did not seem clear in the minds of the people involved in the project. Community leaders were even less clear about tourism development. Some participants expected the project to involve more relatives; some wanted spill-over and to create their own business venture. Some disagreed with the type of management brought from Dili, which was from overseas, thereby keeping locals in the operational roles and not empowered to the managerial role. This structure gave an unwanted ascendant to the NGO personnel and affiliated with figures of authority and expertise that could easily by-pass the community social structure. Lastly, the general sentiment was of over-promises and expectations for different reasons.

The community narratives: rebuilding community and hoping for a future

A lack of business understanding and emphasis on sociocultural aspects were perceived as the root problem of moderate economic success for local communities in achieving their goals of development. The *suku* (community) is a tight-knit group sharing ethnicity, religion, tradition, language and a sense of destiny that was reinforced with the fight for independence. It also has its own lines of fracture: village politics, migrations and the impact of the Indonesian occupation. The barren grey walls of the houses burnt in 1999, and never rebuilt or occupied, are remnants of the Timorese that sided with the occupying forces and did not come back after the independence. In the light of the capacity-building programme, the community authorities appeared to have been engaged at the minimum.

> 'For the completion of the programme, in September, the same act was played. The chief was informed the night before the event. For the first time, I saw also the priest and the police chief. The chief *suku* in the closing ceremony showed a tense expression, and his tone was full of remonstrance in his speech. Despite the attempts by the employee translator to mitigate the speech in the English translation, his embarrassment was strong. The chief had opened the doors to all his grievances: never informed about foreigners being here, not informed of the project, people must be registered, etc. This sense of being snubbed expressed by the chief was received stoically by the participants. They listened but didn't seem affected. It was like the hotel life had created a newly gained freedom from the control of authorities. In the lives of the little community, the beneficiaries of this new environment were satisfied with the change'.
>
> (Research Note, Sept. 2015)

On the one hand, several occasions have shown a pattern of omission to involve the traditional authority (police, chief *suku*, priest). Was it a deliberate omission so as to by-pass authority? It seems that the BFH project offered employees the opportunity to challenge or avoid the traditional power structures. The external power represented through the international team, the international potential clients, the use of English all seem like ways to emancipate the participants of the project – the employees of the hotel. The traditional authorities were never introduced to the group, whether it be the police chief, the chief of the *suku,* or the priest in this deeply Catholic country. As it was a private project, funded by external donors, this omission made sense. However, this disconnect was in direct contradiction to the advertised standpoint of 'community involvement' and the deference for the authorities in place. That was not the case. The traditional local power structure was bypassed, and given minimal recognition.

On the other hand, the national-level power and international experts were put in a position of authority. For both the hotel opening and four months later, for the completion of the programme, the community was invited to participate in the ceremonies. The chief was given due importance as evidenced by his position alongside national politicians, such as the previous president and prime minister of the country. However, it appeared that the chief was informed at the last minute about it. This lack of inclusion was obviously contributing to tensions between community members. The new situation had created a loss of control in the lives of the little community. The local padre followed, giving a blessing ceremony after the official graduation of students. Interestingly the inaugural ceremony of the hotel, a few months earlier, was preceded with a night ceremony conducted by the village sorcerer. Two chickens and a pig were sacrificed outside the main entrance and a tree trunk was planted to commemorate the event, soaked with the animals' blood. This coexistence of traditional beliefs with the markers of the Catholic faith of Timorese was very much a clear sign of the current dynamics of the Timorese identity. The role of the priest's prayers, all kept in the background, is important but also not relevant to the technicality of tourism.

The positioning of the hotel as a catalyst for local development, and the very nature of the capacity-building programme were all geared towards standardized levels of service expected from a 'heritage boutique hotel'. The renovation of the fortress into a boutique hotel brought narratives of economic growth: employment and extra revenue for all. This created confusion about the impact of tourism among the community. For instance, the moto-taxi drivers, who are mostly young men who do little or no business, were expecting tourists to come and use their services. After six months, they felt disappointed to see that tourism brought little benefit for them. The notion of tourism for the community did not really materialize in the minds of people involved in the project. Community leaders were even less clear about tourism development. Local assumptions about tourism, including the expectation of increased income from *malae* (foreign visitors) emerged without any benchmarking or understanding of tourist expectations and without understanding the tourists' profile and their segmentation. The actual objectives were not fully understood, and it would require a specific workshop or training to get a clear understanding and to determine the feeling of exclusion felt by the local community and its leaders, who were ill prepared for the challenges of tourism (Table 12.1).

The tourism narratives of promised economic growth brought by tourism with employment, empowerment and benefits for the community remain challenged. Balibo already had some basic tourism infrastructure: one homestay with four bedrooms and basic features, the possibility of using the moto-taxi, a religious shelter, a community centre with the internet connection and basic communication services. To this extent, Balibo is very similar to other places of Timor Leste that are equipped with similar infrastructure thanks to enterprising locals. This constitutes the

Table 12.1 Stakeholders and the project

Stakeholder	Interest/aspirations	Problem perception	Resources	Role
Participation	• Learn skills • Serve	• Lack of training • Language skills	• Motivation • Employed	• Supportive
Private Operator	• Fulfil tourists' expectations • Revenue	• Limited numbers • Training of staff	• Funding • Management	• Champions
NGOs	• Sustainable development • Success story	• Lack of training • Lack of opportunities	• Funding from donors • Expertise	• Advocates opinion leaders
Local Authorities	• Job creation • Income	• No control	• Police authority	• Marginal
Central Government	• Tourism revenue growth	• Lack of infrastructure • Lack of tourists • Revenue	• Authority • Ownership	• Decision makers
Visitors	• Standards of quality 'comfortable adventure'. • Experience, rugged adventure. • Authenticity	• Little activities available	• Willingness to pay for improved services	• Supportive opinion leaders

backbone of tourism in Timor even if some fail to sustain their businesses in the long term due to the lack of infrastructure and understanding about tourism management. Various projects were brought by NGOs to strengthen ecotourism and CBT infrastructure.

However, the presented project aimed at developing tourism supply through training the staff of a boutique hotel. This project was of national value in introducing a new type of clientele and thus of tourism geared towards heritage and culture. After more than two years, the stream of tourists remains limited, and the only economic benefits are in the areas of tour guides and car rentals. The moto-taxi drivers, mostly young men with little business, have not seem the benefits of tourists arriving. The explosion of tourism has not materialized. R,8 who spoke good English and had lived in Australia, was the most vocal to highlight the issues related to the loss of momentum, and lack of appropriation and empowerment of the community, despite the earlier narratives.

'THH operating running but after 3 years, there're problems: no coordination with the community about the project, no one doing the promotion for culture and history in Balibo, no any good communication between staff and manager and also no any culture activity for the community' (R, 8).

Having built enough confidence, from past experience of tourism in Dili, he decided to start his own business, an ecolodge with tour operating activities, in the outskirts of the village. This personal disappointment cannot hide the fact that the development of tourism infrastructure by the hotel has probably created a spillover effect, or a catalyst in convincing enterprising local people in starting their own tourism-linked activity.

Looking at a possible local development tourism plan

Is the project reproducing the issues familiar with numerous rural destinations? The BFH is an external creation that benefits a middle-class investor (the owner, who lives in Indonesia). But it is also based on a philanthropic approach in which the NGOs have raised funds for the cost of renovation, community projects and training. Also the hotel company is engaged in a moderately ambitious project, with aims to support the local population, while testing the grounds for heritage tourism from a business perspective. BFH/tourism brought about emancipation from the fatalism of agrarian life or migration to the city.

For the employed, it opened the door for the dreams of access to a better life, that is, through consumption and modern comfort. The discourse of community development and progress needs a careful implementation structure, based on an understanding of the local context. Failing to do so leaves the communities weary and suspicious of the responsible tourism discourse if it does not bring positive change in their daily life. Without a clear benefit to the stakeholders, it is difficult to eliminate this feeling of disinterest. Tourism in rural communities may capacitate and empower local people as actors in tourism planning, development and management. However, it is a process through the inter-linkage of sustainable collaboration between government, NGOs, local authorities, entrepreneurs and the community. The private sector has the capacity to be the enabling factor, but there is a requirement of a consensual direction and a strategic approach. The community should see how benefits are distributed throughout the village. The main recommendations from this experience insist on a contextualization of development programmes and the need to insert them into a larger concerted plan over a longer period. In other words, the tourism development approach should be comprehensive in a form of a master plan.

A Local Development Tourism (LDT) approach would recognize the role of small-sized private companies in leading tourism development, and setting the standards of quality (Table 12.2). The role of a local master plan would help in defining the type of future, desired development by the stakeholders while giving a prominent role to the local community to ensure accountability. This LDT would include an executive committee to regulate, stimulate and enable tourism development. That means that the stakeholders must be able to formulate the vision for the future, and find common ground in order to lay out the policies to reach the expected outcomes. The final outcome and the details of implementation are essential for the plan to achieve its objectives. The development of a comprehensive plan for a community is required in order to coordinate

Table 12.2 Planning aspects for Local Development Tourism

Perspective	Characteristics	Enabling Factors
Time	• Long-term planning	• Master plan and Vision
Governance	• Political contexts • Stakeholders	• Transparency committee • Tourism committee • Understanding of power relations • Clear attribution of roles
Economic	• Resources • Entrepreneurs • Banking	• Size of hotels • Availability of micro-loans • Support from donors
Managerial	• Human resources • Competences	• Quality standards/Regulatory • Training programmes, marketing, operational skills
Social	• Jobs • Network	• Responsible tourism training for the community • Relay from returnees
Cultural	• Influences of decisions	• Cultural norms/contexts

independent initiatives sustainably. That can be a strategy for small communities such as Balibo, where the private project is the enabler. Future research could cover the comparison of small communities' governance and success in addressing the various stages of tourism development. Timor-Leste as a tourism destination remains under developed, but a LDT could be a tool that is applicable in rural destinations with a larger number of tourists and limited resources.

Conclusion

This study used a capacity-building programme as a basis for reflecting on the issue of rural governance and responsible tourism. Can tourism be a success story in small developing rural islands? The reflexive analysis derived from the tourism capacity-building project in Timor Leste shows that the programme achieved its training objectives. The hotel employees are equipped with relevant competencies for a boutique hotel. However, the narratives that a private tourism project is an engine for local development and can create spill-over towards new tourism developments appear far-fetched. In creating supply, a new infrastructure provides a basis for tourism development for an isolated rural community. It does not guarantee a virtuous circle of tourism development.

The training brought a better understanding of tourism, and skills development for many of those participants. The community provides different answers to the project, with a gap felt by some members between promises made to the community and the actual reality after six months of operations. It also highlighted latent tensions within the community, which are revived with the arrival of an external project. Indeed, it shows that the new hotel and the capacity-building programme were only a first step towards larger needs in tourism governance training and management skills. In other words, another stage should be a programme involving community members with coordinated efforts to develop tourism in the destination. Despite attempts made by various stakeholders to develop tourism in the rural areas, and the use of the responsible tourism discourse by public agents and NGOs, there is a need for a strategic approach to tourism management. Central to this is the revitalization of the identity and self-esteem of a young nation that had undergone decades of political turmoil. In the process of removing its remaining sociocultural links to Indonesia, Timor Leste will also need to forge an identity that is removed from its imageries with Brazil to find its new place in Asia. Towards this end, the boutique hotel project could assist in this recovery process; hence the strategic approach should cultivate a 'buy in' from the local community by instilling local pride and identity.

The efforts of small-scale private entrepreneurs can benefit from a larger non-CBT project, provided its scale remains small and it keeps a balance between business sustainability and community support. The balance lies in the negotiation between the needs of service quality and the expectations of the local community. As a result, one of the recommendations would be the design of a comprehensive LDT plan. This would place small private ventures as the enablers for economic growth while governance would accompany change in a responsible-tourism direction accepted by the community, with individual ownership and mutually beneficial tourism training.

References

ADB (2015) Timor-Leste: Fact sheet. Manila: Asian Development Bank. See https://www.adb.org/publications/timor-leste-fact-sheet (accessed 5 January 2017).

Belsky, J. (2004) Contributions of qualitative research to understanding the politics of community ecotourism. In J. Phillimore and L. Goodson (eds) *Qualitative Research in Tourism. Ontologies, Epistemologies and Methodologies* (pp. 273–291). London: Routledge.

Botterill, D. (2007) A realist critique of the situated voice in tourism studies. In A. Pritchard, N. Morgan and I. Ateljevic (eds) *The Critical Turn in Tourism Studies* (pp. 121–129). Oxford: Elsevier.

Bouchon, F. and Lew, B.S.L. (2014) Conceptualizing infusion of practices in the touristified city. In *12th APacCHRIE Conference 2014*. Kuala Lumpur, Malaysia. 21–24 May 2014.

Cabasset-Semedo, C. (2009) Thinking about tourism in Timor-Leste in the era of sustainable development. A tourism policy emerging from grass-roots levels. In C. Cabasset-Semedo and F. Durand (eds) *East Timor: How to Build a New Nation in Southeast Asia in the 21st Century*. Bangkok: IRASEC.

Cabasset-Semedo, C. and Durand, F. (2009) *East-Timor. How to build a new nation in Southeast Asia in the 21st century?* IRASEC (Vol. 9). Bangkok: IRASEC.

Castells, M. (2000) *The Rise of the Network Society*. Oxford: Blackwell Publishers.

Cohen, E. and Cohen, S.A. (2012) Current sociological theories and issues in tourism. *Annals of Tourism Research* 39 (4), 2177–2202.

Currie, S. and Turner, L. (2011) Stakeholder collaboration and contestation in tourism planning: the case of Timor-Leste. In *BEST EN Think Tank XIV Politics, Policy and Governance in Sustainable Tourism* (pp. 54–59). Slovenia: Faculty of Economics, University of Ljubljana.

Franklin, A. (2004) Tourism as an ordering: Towards a new ontology of tourism. *Tourist Studies* 4 (3), 277–301.

Government of Timor-Leste (2011) Strategic Development Plan 2011–2030 (GOVDOC). Dili.

Hall, C.M. (2004) Reflexivity and tourism research – Situating myself and/with others. In *Qualitative Research in Tourism – Ontologies, Epistemologies and Methodologies* (pp. 137–155). London: Routledge.

Hannigan, J. (1998) *Fantasy City*. London: Routledge.

Harrison, D. and Hitchcock, M. (eds) (2005) *The Politics of World Heritage: Negotiating. Tourism and Conservation*. Clevedon: Channel View Publications.

Kalsom, K. (2008) Stakeholders' perspectives toward a community-based rural tourism development. *European Journal of Tourism Research* 1 (2), 94–111.

King, B. and Tolkach, D. (2015). Strengthening community-based tourism in a new resource-based island nation: Why and how? *Tourism Management* 48, 386–398. 10.1016/j.tourman.2014.12.013.

Lee, T.H. (2013) Influence analysis of commuity resident support for sustainable tourism development. *Tourism Management* 24, 37–46.

Martínez Gimeno, J. (2007) Selling avant-garde: How Antwerp became a fashion capital (1990–2002). *Urban Studies* 44 (12), 2449–2464.

May, T. and Perry, B. (2011) *Social Research and Reflexivity*. London: Sage.

Moscardo, G. (2011) Exploring social representations of tourism planning: Issues for governance. *Journal of Sustainable Tourism* 19 (4–5), 423–436.

Mura, P. (2016) Tourism, socio-cultural change and issues. In C.M. Hall and S.J. Page (eds) *The Routledge Handbook of Tourism in Asia* (p. 414). New York: Routledge.

Nunkoo, R. and Ramkissoon, H. (2010) Community perceptions of tourism in small island. *Journal of Policy in Tourism, Leisure and Events* 2 (1), 51–65.

Pereiro, X. (2010) Ethnographic research on cultural tourism: An anthropological view. In G. Richards, W. Munsters, E. Binkhorst, M. Castellanos-Verdugo and J. Collins (eds) *Cultural Tourism Research Methods* (pp. 173–187). Wallingford: CABI.

Ren, C., Pritchard, A. and Morgan, N. (2010) Constructing tourism research: A critical inquiry. *Annals of Tourism Research* 37 (4), 885–904.

Riessman, C.K. (2007) *Narrative Methods for the Human Sciences*. London: Sage Publications.

Sakata, H. and Prideaux, B. (2013) An alternative approach to community-based ecotourism: A bottom-up locally initiated non-monetised project in Papua New Guinea. *Journal of Sustainable Tourism* 21 (6), 880–899.

Salazar, N. (2010) From local to global (and back): Towards glocal ethnographies of cultural tourism. In G. Richards, W. Munsters, E. Binkhorst, M. Castellanos-Verdugo and J. Collins (eds) *Cultural Tourism Research Methods* (pp. 188–198). Wallingford: CABI.

Salazar, N. and Graburn, N. (2014) Introduction. In N. Salazar and N. Graburn (eds) *Tourism Imaginaries* (pp. 1–28). New York: Berghahn Books.

Sassen, S. (2007) *Territory, Authority, Rights: From medieval to global assemblages*. Demopolis. New Jersey: Princeton University Press

Scheyvens, R. and Russell, M. (2012) Tourism and poverty alleviation in Fiji: Comparing the impacts of small- and large-scale tourism enterprises. *Journal of Sustainable Tourism* 20 (3), 417–436.

Sin, H.L. and Minca, C. (2014) Touring responsibility. The trouble with 'going local' in community-based tourism in Thailand. *Geoforum* 51: 96–106.

Simpson, M.C. (2008) Progress in tourism management: Community benefit tourism initiatives – a conceptual oxymoron? *Tourism Management* 29 (2008), 1–18.

Tane, N. and Thierheimer, W. (2009) Research about the connection between different rural tourism types and forms. In *4th Aspects and Visions of Applied Economics and Informatics* (pp. 902–906). Debrecen: Hungar

Tolkach, D. (2013) Community-based tourism in Timor-Leste: A collaborative network approach. Victoria University Submitted. See http://vuir.vu.edu.au/24383/ (accessed 4 August 2016).

United Nations Development Programme (UNDP) (2015) *Report on Human Development 2015*. (J. Klugman, Ed.). New York: UNDP.

Wilson, S., Fesenmaier, D.R., Fesenmaier, J. and Van Es, J.C. (2001) Factors for success in rural tourism development. *Journal of Travel Research* 40 (2), 132–138.

Yang, K. and Pandey, S.K. (2011) Further dissection the black box of citizen participation: When does citizen involvement lead to good outcome? *Publ Adm Rev* 71 (6), 880–892.

13 The Constructs of Responsible Rural Tourism Governance for Belum-Temengor Forest Reserve, Malaysia

Joo-Ee Gan and Vikneswaran Nair

Introduction

Tourism is a major contributor to Malaysia's economy, generating RM 84.1 billion in revenue receipts from 25.8 million arrivals in 2018 (Tourism Malaysia, 2018). In 2016, the tourism industry created 639,500 jobs directly, which constituted 4.5% of total employment in Malaysia (World Travel & Tourism Council, 2017). Admittedly, rural tourism is currently not a mainstay of Malaysia's tourism industry. Although the potentials of this segment are recognized, and the Rural Tourism Master Plan 2001 was formulated for its development (Hamzah, 2004), haphazard planning had arguably hampered its growth. Thus, ironically, after almost two decades, the National Transformation Programme's (NTP) Tourism Lab 2.0 still considered homestays (an important aspect of rural tourism) as a 'new area' to be developed (NTP, 2017).

Rural tourism governance in Malaysia is fraught with challenges. The literature shows that weaknesses in the rural tourism governance framework have led to poor waste management (Rahman & Daud, 2011), poor sanitation (Chan & Baum, 2007), pollution (Nasher *et al.*, 2013) and over-development (Peh *et al.*, 2011). Further, there was very little consultation with or the participation of host communities in this segment (Daldeniz & Hampton, 2013; Jaafar & Maideen, 2012; Yacob *et al.*, 2008; Yip *et al.*, 2006).

With Belum-Temengor forest reserve (Belum-Temengor) as the case study, the chapter examines rural tourism governance, especially the institutional obstacles that affect the planning and development of this segment. Through the twin criteria of environmental sustainability and the protection of host community interests, the study seeks to ascertain whether the governance framework facilitates the sustainable development of rural tourism in Belum-Temengor.

The qualitative method was adopted, whereby in-depth interviews with rural tourism stakeholders in Belum-Temengor revealed that eight constructs were indispensable to a responsible rural tourism governance framework namely, lead institution, stakeholder engagement, power sharing, ecological modernization, accountability, communication, empowerment and fairness. Collectively, these constructs constitute the guiding principles vital to address the weaknesses in the current governance framework. The study contributes to rural tourism in Malaysia through the identification of suitable constructs to improve the prevailing rural tourism governance framework.

Background

Governance framework and responsible tourism

A governance framework is the institutional arrangement that affects a particular field. The very idea of 'governance' anticipates the involvement of private actors. 'Government refers to the exercise of public powers by traditional executive state actors, such as central and local government and its agencies' (Fisher *et al.*, 2013: 502). In contrast, 'governance' functions partly through the contribution of private actors. Increasingly, regulatory networks consist of both government agencies and private actors, and there is often interdependence between both sides which leads to the formalization of a collaborative arrangement (Bramwell & Lane, 2000). Such collaborative arrangements constitute the building blocks of a governance framework. Of course, governance implies some conception of accountability, such that private actors are accountable for their actions to society (Peters, 2012).

The concept of responsible tourism was first formulated by Hetzer (1965). The central tenet of responsible tourism is to minimize interference with the natural environment, encourage respect for cultural diversity, maximize the participation of host community in the provision of tourism services, and at the same time, increase tourist satisfaction. Ethical responsibility among operators and ethical tourism consumption are also important elements of responsible tourism (Bramwell & Lane, 2008). The Cape Town Declaration on Responsible Tourism (2002) has galvanized momentum for responsible tourism practices. Today, the concept of sustainable tourism encapsulates the essence of responsible tourism, calling the stakeholders to develop and manage tourism with environmental protection and the socioeconomic wellbeing of the host community in mind (Szymanska, 2013).

Environmental sustainability and governance framework

Many rural tourist destinations are near the hinterland or located within protected areas, thus unsustainable practices in rural tourism can cause environmental degradation (Ghaderi & Henderson, 2012; Ferreira, 2011). Lane (1994) holds the view that a sustainable management system must aim to conserve sensitive areas, balance the demands of conservation and development, stimulate community-based economic growth and preserve the intrinsic features of rural areas. These should be the goals of a responsible rural tourism governance framework.

The environmental attributes of a rural locality can attract tourists, but poorly managed tourism may cause the deterioration of the very environmental attributes that drew tourists to the destination in the first place. Hitchner *et al.* (2009), writing in the context of the Kelabit Highlands of Malaysia and the Kerayan Highlands of Indonesia, speaks of this paradox as a threat to ecotourism. This paradox is equally applicable to rural tourism, in view of the overlapping boundary between these segments.

Tourism development has compromised the environmental integrity of some rural tourism (and ecotourism) destinations in Malaysia. For example, excessive boating activities at the Kilim Karst Geoforest Park, Langkawi Island has led to a higher (than usual) concentration of cadmium in the waters (Nasher *et al.*, 2013). The study of Jaafar and Maideen (2012) on small enterprises in Redang Island and Tioman Island concluded that the development of tourist accommodation has reached its optimum level. To maintain the rural quaintness or ecological uniqueness of these destinations, limits on development must be imposed. Rural tourist destinations are also under threat from non-tourism activities. For example, uncontrolled clearing of forest land for agriculture and illegal logging activities at Lojing Highlands in Kelantan have largely destroyed the touristic value of the locality (Peh *et al.*, 2011).

Host community interests and governance framework

Host communities, like government agencies, businesses and visitors, constitute a category of tourism stakeholder (Nair *et al.*, 2014). Unfortunately, host community participation in rural tourism development is often lacking (Ghaderi & Henderson, 2012; Haukeland, 2011). Ideally, the planning and management of tourist destinations should be a 'tri-sector partnership' between the public sector, tourism enterprises and local residents. However, such collaboration is complex and consultation is either minimal or perceived by the wider community as tokenistic or disingenuous (Hewlett & Edwards, 2013: 45).

Host communities can give valuable input on tourism decision making (Hewlett & Edwards, 2013; Ghaderi & Henderson, 2012; Haukeland,

2011). Engaging the host community in tourism management fosters the community members' sense of stewardship, which is manifested in volunteerism and partnership with other tourism stakeholders (Ferreira, 2011). Further, Article 5 of the Global Code of Ethics for Tourism (UNWTO, 2001) identifies host community involvement in tourism businesses as a means of poverty alleviation.

Currently, most major tourism operators in rural destinations are non-locals. Local-owned tourism businesses are usually small or medium-sized and cater for the budget travellers (Daldeniz & Hampton, 2013; Jaafar & Maideen, 2012; Yacob et al., 2008; Yip et al., 2006). The main causes are the lack of capital, skills and knowledge. This problem is particularly acute where the *orang asli* communities are concerned (Yip et al., 2006).

Fieldwork at Belum-Temengor

Geographical characteristics and current governance framework

Belum-Temengor is part of the Belum-Temengor Forest Complex (BTFC) that encompasses the northern region of Peninsular Malaysia and part of Thailand. Located in the state of Perak, Belum-Temengor spans across 320,000 ha. The damming of several rivers for the construction of a hydro-electric plant resulted in two man-made lakes, i.e. Lake Kenyir and Lake Temengor. This also caused the submergence of some forest areas, consequently the lakes are dotted with small islands that were once mountain peaks. The largest island, Pulau Banding, linked to the nearest town (Gerik) through a highway bridge, serves as the gateway to Belum-Temengor touristic sites. Belum-Temengor is home to rare flora and fauna and endangered species (Wonderful Malaysia, 2017). BTFC is also recognized by Birdlife International as an Important Bird and Biodiversity Area (Birdlife International, 2018). Figure 13.1 shows the location of Belum-Temengor.

Although Belum forest and Temengor forest are geographically proximate, a dual governance framework came to be when 117,500 ha of forest land were declared the Royal Belum State Park (Royal Belum) in 2007 (Schwabe et al., 2015). In terms of conservation, the same agencies are involved in the governance of both Royal Belum and Temengor, namely, the Department of Forestry, the Department of Environment and the Department of Wildlife & National Parks, Peninsular Malaysia (DWNP). These departments come under the Ministry of Natural Resources and Environment.

The inhabitants of Belum-Temengor are the indigenous *orang asli*. The *orang asli* villages near Tiang River, Sara River and Kejar River are rural tourist destinations, Additionally, there are adventure camps and ecotourism trails in Belum-Temengor. The Malays populate the villages located within the Gerik Municipality and derive their livelihoods from agriculture and small businesses.

Figure 13.1 Belum-Temengor forest reserve
Source: Adapted from https://www.cfs-watch.com/

Where rural tourism is concerned, the Ministry of Tourism & Culture (MOTAC) is the lead federal agency. Additionally, Tourism Malaysia Perak, a state agency, promotes the sector within Perak state. The Department of Orang Asli Development (DOAD) is responsible for *orang asli*, regardless of where they inhabit. The Malaysian Armed Forces maintains a presence in Belum-Temengor, for border patrol and where necessary, anti-poaching operations. Non-governmental organizations (NGOs) such as WWF-Malaysia, the Pulau Banding Foundation and Yayasan Emkay contribute towards conservation and host community welfare.

The distinct advantage that Royal Belum has over Temengor is the assignation of the Perak State Parks Corporation (PSPC) as its environmental management unit (Mustafa, 2011). As the findings will show, the PSPC's monitoring and coordinative role significantly enhances the sustainability of the Royal Belum.

Data collection

The qualitative method of semi-structured face-to-face interviews was adopted. Through purposive sampling, the respondents were approached for their knowledge and expertise on tourism governance and Belum-Temengor in general. However, snowball sampling became more prevalent as existing respondents referred us to other potential participants. We prepared three clusters of interview questions that were environment-oriented, host community-oriented or governance-oriented, administered flexibly depending on a respondent's background, experience or expertise.

We interviewed 29 tourism stakeholders consisting of government officials, NGO activists, nature guides, tourism operators and the *orang asli*. Fieldwork was conducted between July 2015 and February 2016, whereby four visits to Belum-Temengor were made. At the respondents' request, three interviews were conducted in government offices in Kuala Lumpur, Ipoh and Gerik, respectively.

A bilingual (English and Bahasa Malaysia) research statement cum consent form was prepared. Every respondent was briefed on the research purpose and his/her signed consent obtained prior to any interview. Most of the respondents were more conversant in Bahasa Malaysia. Thus, only six interviews were conducted in English. All interviews were recorded and transcribed. Interviews conducted in Bahasa Malaysia were translated into English, with special attention to the nuances and subtleties peculiar to Bahasa Malaysia, so that the essence of the respondents' replies were not lost in translation. The authors also kept field notes and took photographic evidence of the localities visited. These supplement the interview data and enhanced the reflexivity necessary in our thematic analysis of the data (Creswell, 2013).

We adopted the three phases of coding as categorized by Strauss and Corbin (1998). During the open coding phase, the transcripts and field notes were analysed for main features supported by the text. During the axial coding phase, the data was analysed again and the salient ideas organized into themes and sub-themes. Lastly, at the selective coding phase, we constructed the theoretical model of the findings from the themes and sub-themes that emerged.

The Challenge of Environmental Sustainability

Coordination between state agencies

There was good coordination between DWNP, the Department of Forestry, PSPC and the Malaysian Armed Forces in conservation and anti-poaching operations. However, such collaborative measures did little to mitigate the effect of licensed logging in Temengor, which destroyed the touristic values of several *orang asli* villages. In contrast, timber harvesting was prohibited in Royal Belum, by virtue of its status as a state park. Nevertheless, it was feared that the unclear demarcation of boundary between Temengor and Royal Belum might subject the latter to illegal logging.

In terms of tourism planning, the coordination between the main stakeholders, namely the PSPC, MOTAC and Tourism Malaysia Perak was sporadic. In the view of Respondent 11 and Respondent 28, Belum-Temengor is primarily a forest reserve, and therefore, tourism activities are merely peripheral. There was a subtle hint that rural tourism (and ecotourism) should not distract the state agencies from environmental protection. Arguably, such a perception arose from the state agencies' tendency to work in silo.

> 'Synergy between agencies, easy to say ... but in truth it is difficult. But you must find ways, and find like-minded people who are not harping on their legal jurisdiction. ... Especially the institutional arrangement [here], largely, it is like a silo because the law provides your boundary – DWNP has its boundary drawn in the law, Forestry Department's boundary is also drawn in the law'. (Respondent 11)

The lack of multi-stakeholders coordination is also evident in solid waste management. The practice is to remove solid waste from the forest sites and dispose of them at the rubbish skip at Pulau Banding. However, solid waste collection is under the jurisdiction of the Gerik Municipal Council, whose collection schedule (twice per week) does not accommodate the higher quantity of waste from tourism activities. Between 2014 and April 2017, when tours departed from a temporary jetty (pending the reconstruction of the main jetty), the infrequent collection often caused a backlog of waste that projected a negative visual impact. This was partly the result of the 'makeshift' waste collection area – a small patch of land without clear signalization of its intended purpose, located at the left of the entrance to the temporary jetty (see Figure 13.2). With the completion of the main jetty in April 2017, a designated waste disposal facility is now

Figure 13.2 Waste disposal area at Pulau Banding temporary jetty in July 2015

Figure 13.3 Waste disposal facility at Pulau Banding main jetty in April 2019

available (see Figure 13.3). Field observations conducted in April 2019 showed that the waste quantity was generally manageable but waste spillage still occurred occasionally. This was because the Council's collection schedule remained the same. This showed that while improved waste disposal facility could reduce waste management challenges, exclusive reliance on Gerik Municipal Council for solid waste disposal – on prevailing terms – could not resolve the problem.

Implementation and enforcement of laws and regulations

There is control on tourist movement in Royal Belum through an army check-point, followed by another PSPC ranger check-point. However, similar control is absent in Temengor. As conservation is the priority in Royal Belum, tourists can only access specific sites within 1.5 km from the water line of the lake. The PSPC rangers are responsible for enforcing this rule, and most licensed nature guides abide by this requirement. In contrast, there is no strict separation between tourism forest sites and other protected areas in Temengor. Solid waste management is better in Royal Belum, due to ranger patrols and the vigilance of the licensed guides. In Temengor, however, unlicensed guides and unaccompanied individuals often litter the forest sites.

So long as Temengor remains largely a production forest, the control mechanisms that exist in Royal Belum cannot be implemented there. In the view of PSPC respondents, Temengor should have its own environmental management unit. Such an agency can only be set up if various government agencies collaborate to overcome their jurisdictional limitations.

Level of voluntary compliance

A high level of self-regulation exists among the nature guides licensed to bring tourists into Royal Belum. This significantly improves solid waste management and the observance of no-go zones in the state park. In Royal Belum, no unaccompanied tourists are allowed. This means that the PSPC can usually pinpoint any transgression to specific licensed guide by checking the record at the entry check-point. According to PSPC respondents, transgressions are few and far between, and invariably resolved through mediation. In contrast, unlicensed guides and unaccompanied tourists can enter Temengor forest sites unhindered. Thus solid waste management is less effective there.

> 'For now ... before the rangers deal with the rubbish, the next guide would have cleared it and report to his own guide association. It is usually like that ... the last team [for the day], they will collect the rubbish. Normally, once the rangers are involved, we will make a black and white report and so on. After that we will give the formal complaint to the association and they will focus on the person we mentioned. So they will block the guide from operating for a certain time frame for not following the rules'. (Respondent 23)

Involvement of NGOs and other private actors in environmental protection

Many NGOs and private actors are involved in the conservation of Belum-Temengor. However, the local NGOs such as the Pulau Banding Foundation and Yayasan Emkay have a more permanent presence there. A visible example of NGO initiative (in collaboration with DOAD) is the construction of solar panels at some tourism villages. The purpose is to power water pump/filtration system or basic electrical appliances. Such projects partly mitigate the harshness of the Belum-Temengor terrain which renders electricity and safe water supply impossible or too costly. Figure 13.4 shows an example of the solar panel used in Belum-Temengor.

Obstacles in the Advancement of Host Communities' Interests

Consultation with the host community

The lack of consultation with the *orang asli* in forestry matters has deprived them of the stewardship of their environment. Such consultation is absent because the forestry law in Malaysia does not mandate consultation with the host community. For example, under section 11 and section 12 of the National Forestry Act 1984, a state government can excise land

Figure 13.4 Solar panel used in Belum-Temengor

for logging without consulting the local inhabitants. As mentioned above, logging activities in Temengor have destroyed tourism in several *orang asli* villages. Other aspects of the *orang asli* livelihoods are also affected. In particular, hunting and gathering in forest reserve are now prohibited, though tolerated by the PSPC and other agencies if pursued for self-consumption. Most *orang asli* abide by this restriction, although there is some harvesting of agarwood for sale to middlemen. The forest reserve status of Belum-Temengor means that agricultural activities are restricted. This has caused hardship to the *orang asli*, particularly since native fish species and other forest produce are being depleted due to logging.

With regard to tourism decision making, there was some consultation with local Malay business operators. However, most respondents agreed that only a small 'inner circle' of operators were consulted. Where tourism planning and development were concerned, consultation with the *orang asli* was almost non-existent.

Host community involvement in rural tourism

In private sector employment, there is no law or policy that requires the hiring of an *orang asli* guide or porter in trekking or camping trips. In public sector employment, there is no active recruitment of *orang asli* as forest rangers, and no formal recognition of their role in conservation. Moreover, there is no systematic tourism capacity-building programme to develop rural tourism in *orang asli* villages.

Overall, tourism businesses in Belum-Temengor employ more local Malays than the *orang asli*. The latter's lack of education and reluctance to commit to full-time employment dissuade tourism businesses from employing them. Most tourism operators, nature guides and boat operators were also local Malays. Some tourism villages produce handicraft for sale at a small scale. Such collective enterprise constitutes the tourism business of the *orang asli*.

Support or assistance from government agencies

There was no entrepreneurial development for tourism operators of Belum-Temengor. The existing nature guide or 'green badge' training was more beneficial to employees compared to tourism operators. Jurisdictional limitation was a problem in devising appropriate assistance to the tourism operators. According to Respondent 29, while Tourism Malaysia Perak actively promotes tourism in Royal Belum, it cannot initiate entrepreneurial training as tourism capacity-building comes under the purview of MOTAC.

Respondents with tourism businesses complained that most tourism operators did not qualify for government-initiated entrepreneurial loans due to the lack of paper qualifications and collaterals. Without capital, these operators could not upgrade their speed boats/house boats. Consequently, many tourism operators could not expand their businesses.

> 'Too many conditions! I can't afford to comply. ... For a loan you need collateral. ... If you want these entrepreneurial schemes, they want many certificates? The certificates ... where do we have? We've been operating here, where are we to find certificates? Our credentials are from the tourists ... they know us'. (Respondent 25)

The role of NGOs and private actors in the empowerment of the host community

The NGOs and private actors provide mainly food-healthcare-education assistance to the *orang asli*. They have yet to initiate tourism capacity-building projects in *orang asli* villages. Yayasan Emkay is an NGO that supports tourism villages in Belum-Temengor. The itineraries of its environmental awareness programmes invariably include visits to *orang asli* villages. However, to facilitate tourism development in these villages, what the *orang asli* need is tourism know-how – this is currently not forthcoming from the NGOs and private actors.

The importance of effective channel of redress

The channel of redress was convoluted and not particularly effective. Locating the right tourism agency to contact was itself a challenge. The tourism operators were generally frustrated with conveying their complaints to local offices, in the hope that their grievances would be brought before the right state or federal agency. Such indirect means of redress lack transparency and accountability. The *orang asli* had not voiced many grievances concerning tourism thus far. In other matters (e.g. forestry), the *orang asli* lacked a channel of redress.

> 'I don't know ... in Perak there's this problem with tourism. That's why we ask the councillor to help and intervene. There are three bodies representing tourism. One is MOTAC, another one Tourism Perak, another ... Tourism ... whatever ... Just who to approach? ... So now we only contact the MP (Member of Parliament). The councillor in charge ... the EXCO (Executive Council) member for tourism in the state. It's easier if we contact him. Whoever he asks to do the job, that's up to him'. (Respondent 26)

Constructs of a Responsible Rural Tourism Governance Framework for Belum-Temengor

The above discussion highlights weaknesses in the current rural tourism governance framework. This led to the identification of eight constructs of responsible rural tourism governance in Belum-Temengor.

Lead institution

The main weakness in the rural tourism governance of Belum-Temengor is the lack of an environmental management unit for Temengor. Whereas the PSPC manages, monitors and coordinates conservation and anti-poaching efforts in Royal Belum; the lack of a comparable agency for Temengor means that such control mechanisms cannot be implemented there. There should be a lead institution in Temengor to coordinate joint operations with DWNP, the Forestry Department, Malaysian Armed Forces, and other agencies. The PSPC simply does not have the mandate or the manpower to undertake the same tasks for Temengor. Yet, in the long run, the environmental integrity of Royal Belum depends on the protection of Temengor forest.

Stakeholder engagement

A responsible rural tourism governance framework should establish constructive relationships among the stakeholders. Such collaboration can be improved upon in terms of frequency, extent of stakeholder involvement and effectiveness. Constructive stakeholder relationships also enhance self-regulation, in that stakeholders who internalize their roles in the fulfilment of a common agenda are more likely to abide by the rules (Fisher *et al.*, 2013).

Stakeholder engagement also includes the establishment of a tourism partnership. As the findings suggest, the Gerik Municipal Council, the nature guides and tourism operators play important roles in the waste management of Belum-Temengor. A tourism partnership should be formed between these parties. It is highly probable that formal recognition through tourism partnership may incentivize Gerik Municipal Council to assume greater responsibility for the waste management of Belum-Temengor.

Power sharing

Currently, the governance framework of Belum-Temengor does not formally recognize the rights of the *orang asli*, or their roles in the stewardship of their own environment. This loophole can be addressed by adopting a power-sharing model such as the Community Use Zone established in Ulu Papar, East Malaysia (Majid Cooke & Vaz, 2011). Importantly, the *orang asli* should be allowed to sustain their livelihoods from specific areas of the forest reserve. This may mean permitting some agricultural activities and the formal recognition of hunting and gathering rights.

Ecological modernization

The concept of ecological modernization states that business profits and growth are attainable without sacrificing the environment (Buttel, 2000). This concept is consistent with sustainable tourism in that environmental protection is non-negotiable. A responsible rural tourism governance framework should embrace ecological modernization in energy management, waste management and the setting of development limits, especially in the construction of tourism facilities and infrastructure. Ecological modernization also requires regular monitoring of the limits of acceptable change (Sofield, 2012), so that rural tourism in Belum-Temengor does not pose a threat to the ecosystem.

Accountability

Currently, rural tourism governance in Belum-Temengor lacks accountability. This is manifested in the lack of consultation between the relevant agencies and the host communities, be they local Malays or *orang asli*. To say that a government agency is accountable means it is answerable to the public for its decisions (Kaltenborn *et al.*, 2011). For accountability to exist, the host community must be provided with sufficient information concerning proposed tourism projects, so that it can participate meaningfully at a consultative forum. The host community should also be given a chance to present its view regarding such projects – there should be a venue for the voicing of opinion, or a mechanism for soliciting responses from the community. At present, consultation only takes place between the relevant agencies (e.g. PSPC and the DWNP) and a small group of tourism operators. Transparency is lacking in that consultation is not broad-based and this worsens the problem of accountability.

Communication

The current channel of redress is convoluted and indirect. The problems with an indirect channel of communication are numerous: the complaints may be 'lost in transmission'; selective filtering of complaints may be practised, such that some grievances never reach the right agency; a complainant may have difficulties following-up with his complaint; a grievance 'conveyed' is not attributable to a specific complainant, hence there is less urgency to address the complaint; and delays in government agency's response – due to some or all of the above problems. An important construct for a responsible tourism governance framework is therefore the establishment of a clear and direct communication channel between tourism operators/*orang asli* and the relevant government agency.

Empowerment

In this context, there are two main facets of 'empowerment' as a construct for responsible rural tourism governance. First, empowerment should include tourism capacity building (Majid Cooke & Vaz, 2011) because the *orang asli* lack skills and knowledge necessary to enter into tourism employment or business. Secondly, tourism operators should be empowered in terms of entrepreneurial support and access to resources, so as to produce 'engaged entrepreneurs' who improve their products/services in response to a competitive business environment (Joppe et al., 2014: 51). The European Unions' LEADER+ is an example of an empowerment programme that aims to improve community capacity in planning and execution of local development initiatives, including tourism projects (Joppe et al., 2014; Strzelecka, 2015). Characteristics of tourism governance pursuant to LEADER+ include the forming of local action groups, the provision of technical expertise and financial support (Strzelecka, 2015). Such empowerment measures are currently lacking in the rural tourism governance of Belum-Temengor.

Fairness

In the context of tourism governance, fairness refers to impartiality. This concept also encompasses 'equity', in that the relevant stakeholders are given the chance to improve their well-being (Kaltenborn et al., 2011). In essence, fairness as a construct puts the democratic character of tourism governance to the test. Fairness in consultation, economic involvement of host community and equitable distribution of tourism benefits can only be achieved where decision-makers are accountable, transparent, engage in dialogue, collaborate, delegate and empower. These are attributes of bottom-up planning – generally considered the preferred governance model for tourism, especially where rural/indigenous communities are concerned (Majid Cooke & Vaz, 2011). As the findings show, fairness is sorely lacking when the host communities in Belum-Temengor have little say in rural tourism development and the *orang asli* barely benefit from rural tourism.

Conclusion

First, the rural tourism governance framework in Belum-Temengor can only be said to adequately promote environmental sustainability. Although the PSPC is reasonably effective in the environmental management of Royal Belum, there is no similar body in Temengor. Consequently, Temengor is plagued by logging activities, unregulated visitor entries and littering. An integrated environmental management of Belum-Temengor calls for the establishment of a lead institution for Temengor.

Secondly, the governance framework does not adequately protect host community interests. Only a small group of local Malays are involved in rural tourism in Belum-Temengor. Very few indigenous *orang asli* are employed in or operate rural tourism businesses. The empowerment of host community members and an equitable power sharing arrangement are especially important to redress these problems.

Thirdly, the distinct institutional arrangements for Royal Belum and Temengor lead to two regulatory regimes. Whereas rural tourism is developed sustainably at Royal Belum, the same cannot be said for Temengor. While the lack of a lead institution for Temengor is the main cause, weaknesses in terms of stakeholder engagement, accountability and communication exacerbate the problem. Conservation of Belum-Temengor must be the over-riding goal – to this end, ecological modernization is an important construct.

Despite a better institutional arrangement for Royal Belum, its fundamental development philosophy is based on the 'island' approach used in the establishment of protected areas that originated from the Western philosophy of creating a binary between nature and humans (Mohamad *et al.*, 2013). In doing so, the indigenous *orang asli* continue to be peripheral to the governance of Royal Belum. The role of the *orang asli* as joint custodians of the Park is restricted by the formal institutional arrangement which lacks a rights-based component that is necessary for the cultural and spiritual values of the indigenous community to be effectively incorporated. Essentially the biocentric approach adopted in designing the institutional arrangement for Royal Belum has succeeded in biodiversity conservation albeit without fully achieving social equity. In contrast the lack of consideration for rights-based approaches has resulted in the over exploitation of the natural resources in Temengor. As a consequence, these contestations are reflected in the rural imageries and tourist image of Belum-Temengor specifically the indigenous *orang asli* as an object of the tourist gaze. Instead of celebrating and showcasing the 'Asianess' of the rural tourism experience (oneness between humans and nature), the governance of Belum-Temengor exemplifies the challenges in managing a protected area based on an inherited colonial model of not seeing nature and humans as a unitary whole.

The qualitative method adopted and the small sample size means that the findings are non-generalizable. Difficulty in procuring interviews with government officials is another limitation. Further, the focus of the research is on rural tourism, thus the findings and recommendations are context-sensitive and may not be applicable to other segments.

A national-level study should be undertaken to appraise the rural tourism governance framework in Malaysia. This can be done through the aggregation of multiple local-level studies with environmental sustainability and the protection of host community interests as criteria of good governance. The findings from such multiple studies can be used to assess the overall relevance of the constructs identified here. This study contributes to the limited and dated literature on rural tourism governance in Malaysia. Further, the constructs recommended can be applied to address the shortcomings of the existing rural tourism governance in Belum-Temengor.

Acknowledgement

Field observations conducted in April 2019 to ascertain the veracity of specific data was funded by the Monash University Malaysia School of Business Research Seed Grant no. B-5-19.

References

Birdlife International (2018) Forests of hope site – Belum-Temengor Forest Complex, Malaysia. See http://www.birdlife.org/worldwide/projects/forests-hope-site-belum-temengor-forest-complex-malaysia (accessed 26 June 2018).

Bramwell, B. and Lane, B. (2008) Priorities in sustainable tourism research. *Journal of Sustainable Tourism* 16 (1), 1–4.

Bramwell, B. and Lane, B. (2000) Collaboration and partnership in tourism planning. In B. Bramwell and B. Lane (eds) *Tourism Collaboration and Partnership: Politics, Practice and Sustainability* (pp. 1–9). Clevedon: Channel View Publications.

Buttel, F.H. (2000) Ecological modernization as social theory. *Geoforum* 31 (1), 57–65.

Cape Town Declaration (2002) The Cape Town Declaration on Responsible Tourism 2002. See http://responsibletourismpartnership.org/cape-town-declaration-on-responsible-tourism/ (accessed 26 June 2018).

Chan, J.K.L. and Baum, T. (2007) Ecotourists' perception of ecotourism experience in Lower Kinabatangan, Sabah, Malaysia. *Journal of Sustainable Tourism* 15 (5), 574–590.

Creswell, J.W. (2013) *Qualitative Inquiry and Research Design: Choosing Among Five Approaches* (3rd edn). Thousand Oaks, CA: Sage.

Daldeniz, B. and Hampton, M.P. (2013) Dive tourism and local communities: Active participation or subject to impacts? Case studies from Malaysia. *International Journal of Tourism Research* 15, 507–520.

Ferreira, S.L. (2011) Balancing people and park: Towards a symbiotic relationship between Cape Town and Table Mountain National Park. *Current Issues in Tourism* 14 (3), 275–293.

Fisher, E., Lange, B. and Scotford, E. (2013) *Environmental Law: Text, Cases and Materials*. London: Oxford University Press.

Ghaderi, Z. and Henderson, J.C. (2012) Sustainable rural tourism in Iran: A perspective from Hawraman Village. *Tourism Management Perspectives* 2–3, 47–54.

Hamzah, A. (2004) Policy and planning of the tourism industry in Malaysia. *Proceedings of the 6th ADRF General Meeting on Policy and Planning of Tourism Product Development in Asian Countries*, Bangkok, Thailand.

Haukeland, J.V. (2011) Tourism stakeholders' perceptions of national park management in Norway. *Journal of Sustainable Tourism* 9 (2), 133–153.

Hetzer, N. (1965) Environment, tourism, culture. LINKS (July): Reprinted in *Ecosphere* 1970, 1(2), 1–3.

Hewlett, D. and Edwards, J. (2013) Beyond prescription: Community engagement in the planning and management of national parks as tourist destinations. *Tourism Planning and Development* 10 (1), 45–63.

Hitchner, S.L., Lapu Apu, F., Tarawe, L., Galih, S. and Yesaya, E. (2009) Community-based transboundary ecotourism in the Heart of Borneo: A case study of the Kelabit Highlands of Malaysia and the Kerayan Highlands of Indonesia. *Journal of Ecotourism* 8 (2), 193–213.

Jaafar, M. and Maideen, S.A. (2012) Ecotourism-related products and activities, and the economic sustainability of small and medium island chalets. *Tourism Management* 33, 683–691.

Joppe, M., Brooker, E. and Thomas, K. (2014) Drivers of innovation in rural tourism: The role of good governance and engaged entrepreneurs. *Journal of Rural and Community Development* 9 (4), 49–63.

Kaltenborn, B.P., Qvenild, M. and Nellemann, C. (2011) Local governance of national parks: The perception of tourism operators in Dovre-Sunndalsfjella National Park, Norway. *NorskGeografiskTidsskrift – Norwegian Journal of Geography* 65, 83–92. Doi: 10.1080/00291951.2011.574320

Lane, B. (1994) What is rural tourism? *Journal of Sustainable Tourism* 2 (1/2), 7–21.

Majid Cooke, F. and Vaz, J. (2011) The Sabah ICCA review: A review of indigenous peoples' and community conserved areas in Sabah. Kota Kinabalu, Malaysia: Global Diversity Foundation. See http://www.researchgate.net/publication/ 272510360 _ The_Sabah_ICCA_Review_Majid_Cooke_and_Vaz_2011 (accessed 2018).

Mohamad, N.H., Kesavan, P., Abdul Razzaq, A.R, Hamzah, A. and Khalifah, Z. (2013) Capacity building: Enabling learning in rural community through partnership. *Procedia – Social and Behavioral Sciences* 93 (21), 1845–1849.

Mustafa, M. (2011) *Environmental Law in Malaysia*. London: Walter Kluwers.

Nair, V., Munikrishnan, U.M., Rajaratnam, S.D. and King, N. (2014) Redefining rural tourism in Malaysia: A conceptual perspective. *Asia Pacific Journal of Tourism Research*, doi: 10.1080/10941665.2014.889026

Nasher, E., Lee, Y.H., Zakaria, Z. and Surif, S. (2013) Assessing the ecological risk of polycyclic aromatic hydrocarbons in sediments at Langkawi Island, Malaysia. *The Scientific World Journal*, doi: 10.1155/2013/858309

National Transformation Programme (NTP) (2017) National Transformation Programme Annual Report 2016. See https://www.pemandu.gov.my/assets/ publications/annual-reports/NTP_AR2016_ENG.pdf (accessed 26 June 2018).

Peh, K.S.H., Soh, M.C.K., Sodhi, N.S., Laurance, W.F., Ong, D.J. and Clements, R. (2011) Up in the clouds: Is sustainable use of tropical Montane cloud forests possible in Malaysia? *Bioscience* 61 (1), 27–38.

Peters, B.G. (2012) Governance as political theory. In D. Levi-Faur (ed.) *Oxford Handbook of Governance* (pp. 49–65). London: Oxford University Press.

Rahman, S.A. and Daud, N. (2011) An analysis of tourists' attitudes towards sustainable tourism: Application to Malaysia. *International Journal of Sustainable Development* 14 (3/4), 206–224.

Schwabe, K.A., Carson, R.T., De Shazo, J.R., Potts, M.D., Reese, A.N. and Vincent, J.R. (2015) Creation of Malaysia's Royal Belum State Park: A case study of conservation in a developing country. *Journal of Environment and Development* 24 (1), 54–81.

Sofield, T.H.B. (2012) Biodiversity conservation and tourism: Are they compatible? Keynote address, *International Conference on Biodiversity and Tourism*. Ipoh, Malaysia, April 2012.

Strauss, A. and Corbin, J. (1998) *Basics of Qualitative Research: Grounded Theory Procedures and Techniques*. California: Sage.

Strzelecka, M. (2015) The prospects for empowerment through local governance for tourism: The LEADER approach. *Journal of Rural and Community Development* 10 (3), 78–97.

Szymanska, E. (2013) Implementation of sustainable tourism concept by the tourists visiting national parks. *Journal of Environmental and Tourism Analyses* 1 (1), 64–79.

Tourism Malaysia (2018) Malaysia tourism statistics in brief. See https://www.tourism.gov.my/statistics (accessed 29 July 2019).

UN World Tourism Organization (UNWTO) (2001) Global code of ethics for tourism. See http://ethics.unwto.org/en/content/full-text-global-code-ethics-tourism (accessed 26 June 2018).

Wonderful Malaysia (2017) Royal Belum State Park. See http://www.wonderfulmalaysia.com/royal-belum-state-park-malaysia.htm (accessed 26 June 2018).

World Travel and Tourism Council (2017) Travel and tourism economic impact 2017 Malaysia. See https://www.wttc.org/-/media/files/reports/economic-impact-research/countries-2017/malaysia2017.pdf (accessed 26 June 2018).

Yacob, M.R., Shuib, A. and Radam, A. (2008) How much does ecotourism development contribute to local communities? An empirical study in a small island. *The Institute of Chartered Financial Analysts of India (ICFAI) Journal of Environmental Economics* 4 (2), 54–67.

Yip, H.W., Mohd, A., Abdul Ghani, A.N. and Emby, Z. (2006) Participation of local operators in ecotourism services delivery: The nature-based tourism development in Pahang National Park, Malaysia. *Anatolia* 17 (2), 313–318.

14 The Quest for an Essentially Asian Form of Rural Tourism

Amran Hamzah, Vikneswaran Nair and Ghazali Musa

Introduction

As in the definition of the main types of tourism, namely ecotourism, urban tourism, heritage tourism and so forth, the definition of rural tourism is equally problematic and contested. Lane (1994) provided one of the earliest definitions of rural tourism as 'tourism that takes place in rural areas'. Subsequently Lane (1994) described the characteristics of rural tourism as being located in remote areas, involving small-scale enterprise, offering wide-open spaces and being closely associated with nature, heritage, and 'traditional' societies and practices. In Asia the final characteristic – traditional' societies and practices – could be contested given the considerable outmigration of rural dwellers to the cities and the ensuing depopulation that had resulted in dysfunctional rural communities (Rigg, 2016), hence blurring the real and perceived image of 'traditional' agrarian societies.

The United Nations (2018) reported that around 54% of Asia's population resides in urban areas, and many of the world's megacities are located in Asia such as Tokyo (37.1 million), Beijing (17.3 million), Shanghai (20.9 million), New Delhi (22.2 million), Karachi (20.7 million) and Jakarta (26.1 million). Although a similar trend in rural–urban migration has also been happening in the West and other parts of the world, the rise in second-home ownership in rural areas has provided a counterbalance, albeit disrupting the traditional community composition and cohesion. In contrast to the West, second-home ownership in most parts of Asia is still in its infancy. Instead Japan and many countries in Southeast Asia have experienced massive rural depopulation resulting in precarity in the lives of the remaining elderly population (Rigg, 2016). As the urban population grows, the need to escape to tranquil and idyllic settings for recreation has become more pronounced in Asia. In response to this, rural tourism has blossomed not only in the form of tourism that is essentially 'rural' in characteristics but also a spectrum of tourism that takes place in rural areas (Hall *et al.*, 2003). This tourism overlaps with ecotourism, adventure tourism and wellness tourism, etc. and includes offerings and activities such as caving, zip lining, mountain biking, health farms and 'mimicking nature' attractions. Together with this wide spectrum of rural tourism come not only contestations and challenges but also prospects and opportunities that have been amplified in the chapters of this book.

The authors of the chapters in this book were given two main tasks; firstly, to present evidence to support the notion that there is an essentially Asian form of rural tourism. The second task was to debate whether rural tourism in Asia has fulfilled or is aligned to the principles of responsible tourism. Upon reflection, the first task is understandably formidable given that Asia is huge and plural. As such what constitutes 'Asianess' in rural tourism needs to be fundamentally examined by tracing Asia's distinct development philosophy that underpins its development trajectory. At a conference on protected area management in Asia in 2011, which was held in Sendai, Japan and attended by around 20 Asian countries, there was a claim put forward by several of the keynote speakers that Asia has a distinct development philosophy (AIU/MOEJ, 2011). However, such an assertion, at that time, was clouded by confusion and rhetoric but intrigued by this prospect, the International Union for Conservation of Nature's (IUCN) Asia Regional Office subsequently commissioned a study on the Asian Philosophy of Protected Areas (Hamzah, 2015).

The research concluded that harmony between humans and nature had provided Asia with a common philosophical foundation that had shaped its adaptive form of spatial development (Hamzah *et al.*, 2015). Furthermore, the IUCN research surmised that traditionally protected areas in Asia, through taboos, beliefs and prohibitions, had given protection to sacred natural sites such as forests, mountains, rivers and caves. In turn, the reverence for sacred natural sites in tandem with the local society's cosmological worldview had shaped Asia's 'rurality' that is manifested in the region's socioecological production landscape such as the Japanese *Satoyama*. Central to *Satoyama* is oneness between humans and nature and the critical role of humans in continuously adapting to and shaping pristine natural areas to create a unique cultural landscape. Paradoxically while rural depopulation has become a threat in sustaining the *Satoyoma* landscape in Japan, the concept is growing in popularity across Asia due to the advocacy and outreach by the International

Partnership for the *Satoyama* Initiative (IPSI), but more so because the concept resonates well with the worldview and cosmology of traditional societies across Asia.

As opposed to the wilderness associated with national parks in North America and the quaint and arguably fossilized villages and hamlets that dominate the English countryside (Hall *et al.,* 2003), 'rurality' in Asia does not create a barrier between humans and nature, as in the former, nor does it restrict the continuous adaptation of the physical environment to modify nature. While it is true that the North American concept of wilderness is evident in the protected areas across Asia, they are the legacy of a 'colonial model' that is increasingly being contested especially by the indigenous communities who used to live within the Park boundary before they were relocated (Hamzah *et al.*, 2015).

A Framework for Responsible Rural Tourism in Asia

Central to the deliberations in the 12 case-study chapters of this book is the notion of responsible rural tourism. Although the term 'responsible tourism' has been used interchangeably with sustainable tourism, there are fundamental differences as highlighted by Nair *et al.* (Chapter 1). Sustainable tourism has been subjected to endless scrutiny by scholars and practitioners, and both its definition and operationalization have been described as 'ambiguous', 'an oxymoron', 'a paradox', 'inconsistent', etc. (Butler, 2015). Nonetheless, Ruhanen *et al.* (2015) conducted a bibliometric analysis of 492 published papers over a 25-year period to reveal a substantial body of knowledge on sustainable tourism that has progressed from definitional and conceptual papers to testing and applying theory through empirical papers.

Responsible tourism, on the other hand, could be viewed as a form of backlash to the inherently problematic conceptualization and delivery of sustainable tourism, especially from practitioners by offering a pragmatic and implementable set of guiding principles to make travel more benign and ethical. The call and campaign for responsible tourism started in the UK in the mid-nineties and South Africa was the first country to adopt responsible tourism as a national policy in 1996 (Goodwin, 2007). In 2002, the Cape Town Declaration on Responsible Tourism was launched which provided practical principles for the tourism industry to be more responsible and ethical. It is not the intention of this section to revisit the definition and interpretations of responsible tourism. However, it is necessary to prelude the development of a framework for responsible tourism in Asia by placing it in a universal context that is in line with the practical definition of responsible tourism, which is 'to create better places' (Goodwin, 2007), as outlined in the following chart Figure 14.1.

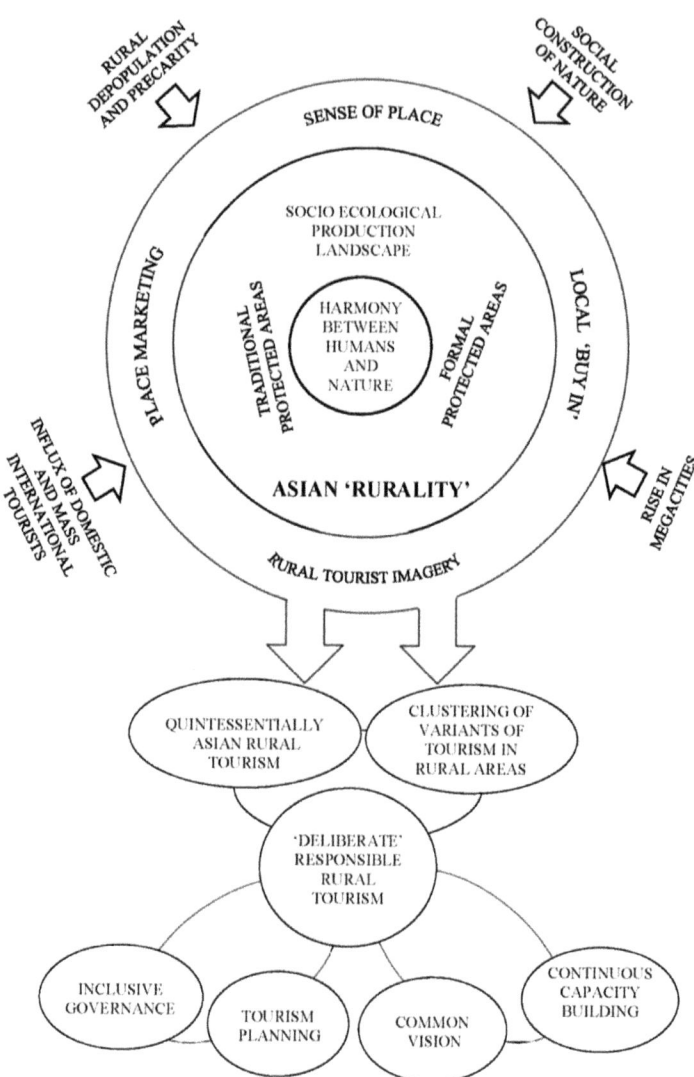

Figure 14.1 Framework for responsible rural tourism in Asia

The Underlying Asian Development Philosophy as a Guiding Principle

An Asian framework of responsible rural tourism has to start with the adherence to a development philosophy of oneness with nature. The various chapters in this book have revealed that the Asian philosophy had been manifested in the socioecological production landscape of Asia and is similar to *Satoyama*, which in turn, has created the sense of place and tourist imagery for intrinsically Asian forms of rural tourism to flourish. In addition, 'formal' protected areas such as national parks have added another dimension to Asian rural tourism notwithstanding the fact that they are the by-products of a colonial legacy. Overlapping with ecotourism and adventure tourism, national parks have provided the setting for the development of activities that are compatible with Asia's social construction of nature such as trekking, mountaineering, caving and so forth. However, a spiritual element needs to be added and emphasized to differentiate Asian national parks based on the reverence for sacred natural sites such as mountains and forests, etc. The case of Kinabalu Park in Sabah, Malaysia illustrates this peculiarity in which the reverence for the sacred mountain had been the main reason for its protection by the local community (van der Ent, 2013). As a result of the ignorance of the spiritual value of Mount Kinabalu, the infamous nude photo opportunity (and its aftermath) displayed by a small group of Western tourists at the mountain peak in 2015 (*The Star*, 5 June 2015) illustrated not only the incompatibility of hedonistic behaviour with the principles of responsible rural tourism but also the Asian regard for mountains, forests and caves, etc. as a biocultural rather than a natural phenomenon (Hamzah, 2017). As such, it is imperative for responsible rural tourism in Asia to manifest the philosophy of harmony between humans and nature as the foundation for its spatial development.

Linking Cultural Landscape with the Rural Tourist Imagery

Despite its challenges, the role of tourism as a catalyst for rural revitalization has to be accepted as a responsibility given the 'boosterist' (Hall, 2008) attitude taken by governments in relation to the economic potential of tourism. The limited success of community-based tourism (CBT) in fulfilling this role (Goodwin & Santilli, 2009) should be viewed not as the inherent weakness of this approach but more so a result of a lack of systematic planning and implementation. Nevertheless, whether tourism could be regarded as a panacea to tackle the more challenging, and particularly Asian issues of depopulation and rural precarity remains to be answered.

From a place marketing perspective, the quintessentially Asian forms of rural tourism presented in this book have demonstrated how a distinctive tourist imagery could be formed by the local resources and sense of place. Beyond the success stories of tea tourism, monsoon tourism and wellness tourism presented in this book, the iconic Banaue rice terraces in the Philippines and the equally majestic Tegalalang rice terraces in Bali are compelling examples of how Asia's socioecological production landscape has played a central role in the formation of the rural tourist imagery. In reality though, the encroachment by tourism development in the form of villas mainly owned by foreigners are threatening not only the authenticity and integrity of the rice terraces in Bali but the more important *Subak* water temple system that has survived for more than 1000 years and is still integral to community live and cohesion. The fact that both the WH-listed cultural landscapes have been 'yellow flagged' by United Nations Educational, Scientific and Cultural Organization (UNESCO) in the past is a reflection of their fragility in facing constant threats from unsustainable development. This amplifies the crucial role of responsible rural tourism in biocultural conservation in Asia that goes beyond preserving the rural idyllic and extends into maintaining sustainable rural livelihoods. Suffice to say that sustaining the traditional cultural landscape that supports sustainable rural livelihoods is a critical factor in developing a framework for responsible rural tourism in Asia.

An Endogenous Governance Structure that Incorporates the Traditional Power Structure of Asian Societies

The case studies presented in this book also share a commonality in highlighting the critical role of governance in paving the path for the creation of responsible rural tourism. While the prominent role of the government is still evident especially in the case of institutionalized agritourism as in the case of the Philippines, its effectiveness in ensuring local 'buy in' and a long-term commitment to responsible rural tourism remains to be seen. In contrast the various case studies related to CBT have shown a greater focus on participatory governance rather than government (Goodwin, 1998). Government refers to the formal institutional structure and authority, whereas governance refers to a more inclusive distribution of power in the decision-making structure that is representative of the key stakeholders such as the government, industry players, local community and non-governmental organizations (NGOs) (Sharpley, 2003). As evident in the CBT chapters in this book, an endogenous governance system that incorporates the traditional power structure is more effective in developing a common vision right from the onset of the project in line with the principles of responsible rural tourism. Central to an inclusive governance structure is the incorporation of Asian values such as kinship and ethnical homogeneity as well as having a clear channel of communication for native and customary laws to be integrated into the formal decision-making process. For most rural communities that are lacking in local capacity, the advent of tourism is alien to their worldview, let alone having to adhere to the principles of responsible tourism. Although capacity building is often used as a solution to overcome the problem of lack of local capacity, the focus has always been on superficial skills advancement. What is required, in support of a framework for responsible rural tourism, is a common vision and a governance structure that is able to ensure local empowerment, control and benefits, as rural areas and 'rurality' in Asia are being increasingly commodified as tourism offerings.

'Deliberate' Responsible Rural Tourism

Due to the convergence of ecotourism, adventure tourism, wellness tourism etc. in rural areas, it has to be accepted from the practical perspective at least, that tourism that takes place in rural areas has to be recognized as a form of rural tourism. By the same token, it is becoming increasingly difficult to distinguish the types of tourism offerings that are incompatible with the rural idyllic and imagery except those that are associated with hedonistic behaviour (Sorenson & Nilsson, 2003). Adding to this haziness is whether hedonistic tourist behaviour is only associated with drunkenness and drug taking or is being loud while travelling in a large group included, given the recent influx of tourists to tourism attractions that are located in the rural areas of Asia?

As demonstrated in the third theme of this book, rural areas in Asia have been engulfed by mainly 'circumstantial' responsible rural tourism for economic reasons. The final factor in formulating a framework for responsible rural tourism is a tourism planning process that is aimed at reducing the fragmentation of tourism development in rural and peri-urban areas so as to create economies of scale. Given that tourism development in rural areas is inherently deficient in economies of scale, it should be developed by optimizing the economies of scope in the form of tourism micro-clusters or thematic rural tourism corridors so as to be able to leverage on business collaboration, cost sharing and joint marketing/promotion (Michael, 2007). Towards this end, micro-clustering can also contribute to destination branding. Most importantly, the tourism planning process should embark on the creation of 'deliberate' responsible rural tourism from the perspective of the industry, tourists and local communities.

Conclusion

The book began by introducing the definition, conceptualization and interpretations of 'rurality' and rural tourism. Given the overlaps between rural tourism and other forms of tourism such as agritourism, ecotourism, adventure tourism etc., agreeing to a universal definition of rural tourism is problematic to the extent that it is relatively easier to define what is not rural tourism.

Subsequently the chapters in the book present detailed case studies of the practice of rural tourism in Asia which overlaps mainly with agritourism and CBT. In answering the question whether there is an intrinsically Asian form of tourism, many of the chapters highlight harmony between humans and nature as the underpinning philosophy that had shaped its cultural landscape and sense of place. In turn, savvy place marketing had built on the uniqueness of Asia's 'rurality' to portray and sell a tourist imagery that is distinctively Asian. In contrast to the rural areas in

Western countries, the holiday home syndrome has yet to become a major phenomenon in Asia's rural areas in comparison to rural depopulation and precarity. In response to this peculiarity, all chapters in this book either explicitly or implicitly recognize the role of rural tourism as a panacea to both the problems of economic underdevelopment and rural depopulation. While it is not certain whether rural tourism could effectively play this challenging role, there are ample lessons to be learned from the case studies to suggest that responsible rural tourism in Asia can only be achieved by building on its ancient philosophy and an endogenous governance structure that not only recognizes the traditional power structure but is also flexible and nimble enough to accommodate contemporary tourist demand and changes, especially from the ever-increasing urban population that has migrated to the megacities in Asia.

As mentioned earlier Asia is vast and diverse, and except for a common ancient philosophy, the region offers extreme variants of rural tourism. The chapters focus mainly on the supply side of rural tourism. Thus, there is a large research gap in relation to the demand for rural tourism especially from the rising urban population of Asia. In relation to this, further research is required on the social construction of nature from the perspective of domestic tourists. Modernity, the internet, mainstream religions and a globalized world (Hamzah *et al.*, 2015) are threatening the spiritual relationship between contemporary Asian societies with nature, which in turn, is being manifested in their social construction of nature, travel motivation and behaviour. In addition, further research from the perspective of industry players is imperative. Given that responsible tourism was initiated by tour operators in the UK, in efforts to offer a more ethical and benign reason for travel, to what extent their counterparts in Asia also subscribe to the same aspiration needs more research.

In an ideal world, the way forward for rural tourism in Asia is exciting and full of opportunities to celebrate and share Asia's contribution to the world, in line with the so-called Asian Century. The quest for an essentially Asian form of tourism is expected to be alive given the rising number of middle class in the dense Asian cities, and their longing for nostalgia and *Furusato*. Discerning Western tourists in the search for the 'Other' will also find satisfaction in the intrinsically Asian tourist experiences that are offered by the tea villages and Kerala backwaters tourism, etc. In response to this type of travel motivation and expectations, savvy tour operators are expected to continue leveraging on Asia's rural imagery as an essential element of place marketing to sell low-volume, high-value rural tourism packages. In the same light, local communities will be content to play their role as joint custodians of the natural resources and continue to preserve their culture as long as they could enjoy a piece of the rural tourism pie.

However, new arrivals to the cities especially the working classes will find more pleasure in the 'family fun' attractions that are mushrooming in the peri-urban areas, and will be relatively less seduced by the rural imagery they had recently left. More importantly their social construction of nature would likely be a 'place to be enjoyed with family and friends' rather than tranquility and wilderness. Moreover, there could also be a growing disconnect with the ancient Asian philosophy of harmony between humans and nature, which in turn, could lead to irresponsible tourist behaviour. How Asia responds to these challenges, to sustain a responsible form of rural tourism, will crucially depend on awareness, responsibility and commitment of the key stakeholders as well as systematic and inclusive tourism planning. Otherwise rural tourism in Asia will follow a trajectory that merely feeds on the rapidly increasing appetite among its ever-growing urban dwellers for tourism that takes place in rural areas.

References

AIU/MOEJ (Akita International University/Ministry of Environment Japan) (2011) *Proceedings of the Post CBD COP 10 International Workshop on Governance in Asian Protected Areas*, Akita International University.
Butler, R. (2015). Sustainable tourism: Paradoxes, inconsistencies and a way forward. In M. Hughes, D. Weaver and C. Pforr (eds) *The Practice of Sustainable Tourism: Resolving the Paradox* (pp. 66–80). London: Routledge.
Goodwin, H. (2007) Advances in responsible tourism. *ICRT Occasional Paper* No. 8, International Centre for Responsible Tourism, Leeds Metropolitan University, UK.
Goodwin, H. and Santilli, R. (2009) *Community-based Tourism: A Success?* German Development Agency.
Goodwin, M. (1998) The governance of rural areas: Some merging research issues and agendas. *Journal of Rural Studies* 14 (1), 5–12.
Hall, C.M. (2008) *Tourism Planning: Policies, Processes and Relationships*. Essex: Pearson Education.
Hall, D., Roberts, L. and Mitchel, M. (2003) (eds) *New Directions in Rural Tourism*. Aldershot: Ashgate Publishing.
Hamzah, A. (2017) World heritage and rights in Malaysia: A case study of Kinabalu Park World Heritage Site, Sabah. In P.B. Larsen (ed.) *World Heritage and Human Rights: Lessons from the Asia Pacific and Global Arena*. London: Earthscan.
Hamzah, A. (2015) Revisiting the Asian philosophy of protected areas. See https://www.iucn.org/content/guest-editorial-revisiting-asian-philosophy-protected-areas (accessed 13 June 2019).
Hamzah, A., Ong, D.J. and Pampanga, D. (2015) Asian Philosophy of Protected Areas. Centre for Innovative Planning and Development, Universiti Teknologi Malaysia/ IUCN Asia Regional Office, Bangkok.
Japan Times (2013) *Japan's Depopulation Time Bomb*, Editorial, April 17, 2013.
Lane, B. (1994) What is rural tourism. *Journal of Sustainable Tourism* 2 (1–2),7–21.
Michael, E.J. (2007) Micro-clusters and networks: The growth of tourism. *Advances in Tourism Research Series* 2007.
Rigg, J. (2016) From vulnerability to precarity: Rural lives and livelihoods under conditions of neo-liberalism. Keynote address presented at *RRPG 7th International Conference*, 15–17 August, UniversitiTeknologi Malaysia, Johor, Malaysia.
Ruhanen, L., Weiler, B., Moyle, B.D. and McLennan, C.J. (2015) Trends and patterns in sustainable tourism research: A 25-year bibliometric analysis. *Journal of Sustainable Tourism* 23 (4), 517–535.

Sharpley, R. (2003) Rural tourism and sustainability – A critique. In D. Hall, L. Roberts, and M. Mitchell (eds) *New Directions in Rural Tourism*. Aldershot: Ashgate Publishing.

Sorenson, A. and Nilsson, P. (2003) What is managed when managing rural tourism? The case of Denmark. In D. Hall, L. Roberts and M. Mitchell (2003) (eds) *New Directions in Rural Tourism*. Aldershot: Ashgate Publishing.

The Star (2015) Sabah quake: Mount Kinabalu may be 'angry' with nudists. 6 June 2015. See https://www.thestar.com.my/news/nation/2015/06/06/ mount-kinabalu-may-be-angry-with-nudists/ (accessed 24 January 2019).

United Nations (2018) *World Urbanization Prospects: The 2018 Revision*. Economic and Social Affairs, United Nations.

Van der Ent, A. (2013) *Kinabalu*. Borneo: Natural History Publications.

Index

Note: References in *italics* are to figures, those in **bold** to tables

accessibility 37
agritourism 103–105
agritourism in the Philippines 16–17, 105, 118
   categories of sites **108**, 108–109
   Costales Nature Farms 105, 109–111, 112
   defined 106
   development of 106–108
   methodology 106–109
   responsible rural tourism 103
   conclusion 111–113
American Forests 154
Anderson, V. 5
ARCUS Foundation 154
Ashley, C. 146, 154
Asia Pacific Economic Cooperation (APEC) 144
Asia: urbanisation 213–214
Asian Development Bank 50, 51, 164
Asian forms of rural tourism 13–15, 213–215
   Asian development philosophy 216–217
   cultural landscape and rural tourist imagery 217–218
   'deliberate' responsible rural tourism 219
   framework for responsible rural tourism 215, *216*
   governance structure 218
   conclusion 219–221
   see also Hoi An community organic vegetable gardens; Kerala: responsible tourism; tea tourism in Japan
Australia
   Adelaide Botanic Gardens 81–82
   rural areas **9**
   Undaralava tube caves, Queensland 42

Austria: rural areas **9**
Azmi, R. 4

Bangladesh: monsoon tourism 14, 85
Barrow, E. 129
Belsky, J. 182
Belum-Temengor forest reserve, Malaysia 18–19, 195–196, *199*
   accountability 207–208
   communication 208
   constructs of responsible rural tourism governance 206–209
   coordination between state agencies 200–202, *201*, *202*
   data collection 199–200
   ecological modernization 207
   empowerment 208
   environmental protection 203, *204*
   environmental sustainability 197, 200–203
   fairness 208–209
   fieldwork 198–200
   geographical characteristics 198–199
   governance framework 196–199
   host community interests 197–198, 203, 204–206
   laws and regulations 202–203
   lead institution 206
   NGOs and other private actors 203, 205
   orang asli 19, 198, 200, 203, 204–205, 206, 207, 208, 209, 210
   Perak State Parks Corporation (PSPC) 199, 200–201, *202*
   power sharing 207
   responsible tourism 196
   stakeholder engagement 206–207
   voluntary compliance 203
   conclusion 209–210
Benfield, R. 77

Bernardo, D. et al. 106
Bhutan 11–12, 27–28
　accessibility 37
　adventure tourism 33–34
　Cape Town Declaration 12, 33, **34**, 35, 37
　Clean Bhutan 34–35, 36
　community involvement 35–36
　cultural heritage 36–37
　GNH (gross national happiness) 28, 29, 30–31, 33, **34**, 37, 38
　Jigme Dorji National Park 35
　minimizing impacts 33–35
　mountains 35
　nature conservation 36
　Royal Manas National Park 36
　rurality 28
　sensitivity, pride, respect 37–38
　sustainability 28, 29–30, 31–32
　tourism 28–30, 31–32
　Tourism Council 30
　trekking 30, 33–35
　conclusion 38–39
biocentricity 41, 42
Blackstock, K. 117
Botterill, D. 183
Bramwell, B. 6
Brundtland report (1987) 3, 32, 88
Brunet, S. et al. 31
Buddhism 44, 45, 46, 47
Butler, R. 215

Cabasset-Semedo, C. 179, 181
Cambodia
　community-based ecotourism (CBET) 164, 166
　Complementary Approaches 174
　geography and economy 162–163
　government tourism policy 163–164
　I/NGOs 164, 165
　'MlupBaitong' 165–167, 169, 170, 171, *172*, 174–175
　see also Chambok community-based ecotourism
Cambodian Community Based Ecotourism Network (CCBEN) 165
Cambodian Tourism Strategic Directions Plan (2016–2018) 164, **165**
Canada: rural areas 9

capacity building 15, 17, 91, 131, 145, 218
　see also Chambok community-based ecotourism, Cambodia; Miso Walai Homestay, Sabah, Malaysia; Timor Leste rural tourism
Cape Town Declaration on Responsible Tourism (2002) 5, 33, **34**, 56, 88–89, 103, 173, 196, 215
　see also Bhutan
cave tourism in China 12–13, 41, 43–44
　Buddhism 44, 45, 46, 47
　Confucianism 44, 46
　Crown Cave, Guilin *48*, 48–50, *50*, *51*, 52–54, **54**, 55–57
　Daoism 41, 44, 45–46, 47
　dragons 47–48
　elements of physical universe 50, *51*
　Hubei Province 44, 45, 46–47
　'mainstreaming' for poverty alleviation 50, 51–54, **54**, 57
　nature and naming of caves 44–48
　responsible rural tourism 41
　Three Gorges 46, 47
　Wudangshan 44
　Yaolin Cave, Hangzhou 47
　conclusion 55–57
caves
　classification **43**
　western management of 41, 42–43
　see also cave tourism in China
Cawley, M. 63
CBET (community-based ecotourism) 164, 166
CBT see community-based tourism
CBtT see Communities Benefitting through Tourism
Chambok community-based ecotourism, Cambodia 20, 166, 167, 167–168, *168*
　Cambodia Mine Action Committee (CMAC) 167
　Community Fund 169
　early difficulties 169–170
　Ecotourism Management Committee (EMC) 166
　Kirirom National Park 165, 166
　sustainability *170*, 170–173, *171*
　Women's Forest Café 168, 170, *170*

Women's Self-Help Groups (WSHGs) 168, 173
conclusions: responsible, rural, community-based ecotourism 173–175
China
  intra-Asian tourism 12
  tea tourism 64
  see also cave tourism in China
CIT (community-involved tourism) 10
Cohen, E. 179
Cohen, S.A. 179
Commonwealth Secretariat 51
Communities Benefitting through Tourism (CBtT) 50–54, 55, 57
community-based ecotourism (CBET) 164, 166
community-based tourism (CBT) 10, 19–20, 52, 55, 56, 128–129, 142, 180
  community characteristics 129–130
  community cohesion 17, 18, 126, 129–130
  governance 19, 218
  objectives 116–117
  poverty alleviation 50–52
  social capital 129
  success criteria 129, 217
  see also community-based tourism in Vietnam; Coruh Valley, Turkey; Miso Walai Homestay, Sabah, Malaysia
community-based tourism in Vietnam 18, 128, 130
  CBT and social cohesion 136–138
  community characteristics 133–134
  embedment of community members 134–136
  methodology 133, **133**
  My Son CBT village 130–131, 132, 133–134, 135, 136, 137, 138
  NGOs 128
  resource management perspective 130
  Triem Tay CBT village 131–132, 133, 134, 135–136, 137–138, 139
  findings 133–138
  conclusion 138–139
community cohesion 17, 18, 126, 129–130
community-involved tourism (CIT) 10
Confucianism 44, 46

Corbin, J. 200
Coruh Valley, Turkey 17
  Eastern Anatolia Tourism Development Project 117–124
  first phase (2007–2009) 118–119
  second phase (2010–2012) 119
  methods 119–121
  results 121–124
  conclusion 124–126
Costales, Ronald and Josephine 110
Crilley, G. 81–82
Currie, S. 181

Daoism 41, 44, 45–46, 47
Denmark: rural areas **9**
developing countries 8, 20, 117, 142, 178, 179–180
disaster risk reduction (DRR) 14–15
disaster tourism 14

Eber, S. 88
ecological modernization 207
economic development 3
environment 55
European Union
  LEADER+ 208
  rural tourism defined 9
experiential tourism 76–78

fairness 208–209
Farrelly, T.A. 129, 130, 138–139
Fiji 130, 138–139
Fisher, A. 64
Fisher, E. *et al.* 196
France: rural areas **9**
Francis, J. 141
Franklin, A. 179

gardens and experiential tourism 76–78
  see also Hoi An community organic vegetable gardens
Gillieson, D. **43**, 48
Gillmor, D.A. 63
glocal ethnography 182
Goodwin, H. 33, 38, 56, 117, 141, 173
governance 18, 196, 218
government, defined 196
Graham, J. *et al.* 18
Greater Mekong Sub-region Tourism Development Program (2005–2008) 164

Greater Mekong Sub-region Tourism Infrastructure for Inclusive Growth Project (2014–2018) 164

Hall, C.M. 1
Hall, D. *et al.* 13
Hamzah, A. 144, 147
Handbook on Community Based Tourism 144
harmony between humans and nature 11–13
  see also Bhutan; China
Hatton, M.J. 147
Hetzer, N. 196
Hitchner, S.L. *et al.* 197
Hiwasaki, L. 116
Hoi An community organic vegetable gardens 13–14, 15, 76, 78–79, 80, 132
  methods 79–80
  nostalgia and sense of place 82
  off-peak tourism rainy season 82–84
  repeat visitors 81–82
  conclusion 84–85
Holland, J. *et al.* 7

Icamina, P. 106
India **9**, 12
  see also Kerala: responsible tourism
India Today 90
Indonesia
  rural areas **9**
  Subak water temple system, Bali 218
  Tegalalang rice terraces, Bali 217–218
International Conference on Responsible Tourism in Destinations (ICRTD) 5
International Labour Organisation (ILO) 131, 132
International Partnership, *Satoyama* Initiative (IPSI) 214–215
International Union for Conservation of Nature (IUCN) 214
International Year of Sustainable Tourism (2017) 1

Jaafar, S. Maideen, S.A. 197
Japan
  AIU/MOEJ (Akita International University) 214
  disaster risk reduction 14–15
  green tourism (*Furusato*) 64, 71
  rural depopulation 213–214
  Sanriku Fukko National Park 14–15
  Satoyama 72, 214–215, 217
  see also tea tourism in Japan
Jurowski, C. 2

Kerala: responsible tourism 15, 89–91
  ABC 89, 98
  Department of Tourism 91
  Destination Level Responsible Tourism Committee (DLRTC) 15
  economic impacts in local community 92, 93–95
  environmental impacts in local community 96–97
  hotel industry links 92, 93, **94**
  Kumily study area and research methods 91, **92**
  major tourist destinations *90*
  NGOs 97
  positive impacts of initiatives 92
  Responsible Tourism initiatives 90–91, *91*
  socio-cultural impacts 95–96
  spice shop 94
  State Level Responsible Tourism Committee (SLRTC) 15
  Thekkady village life experience 94–95
  wellness tourism 15
  conclusion 97–98
Khalifah, Z. 144, 147
Kibicho, W. 117
King, B. 180
Kirk, W. 55
Korea: Global Green Growth Institute 97
Kurosawa, A. 14

Lane, B. 2, 6, 7, 13, 197, 213
Lapeyre, R. 116
Li, F.M.S. 46, 48, 52
Liu, Z. 125

McIntosh, A.J. 77
McKercher, B. 82, 125
Malaysia
  Community Use Zone, Ulu Papar 207
  Cooperative Commission (SKM) 143, 153, 155–156
  Corridor of Life project 154–155

Department of Orang Asli Development (DOAD) 199, 203
Desa Lestari programme 155–156, *156*
Kilim Karst Geoforest Park, Langkawi Island 197
Kinabalu Park, Sabah 217
Lojing Highlands, Kelantan 197
Ministry of Tourism & Culture (MOTAC) 199, 200–201, 205
monsoon tourism 14
National Forestry Act (1984) 203, 204
National Transformation Programme (NTP) 195
Pulau Banding Foundation 199, 203
Redant Island 197
reforestation 154–155
rural areas 9, **9**
rural tourism governance 195
Rural Tourism Master Plan 2001 195
Sabah Forest Department (SFD) 149, 154
Tioman Island 197
tourism 195
Tourism Malaysia Perak 199, 200–201, 205
Tungog Rainforest Eco Camp (TREC) 149–151, *150*, 157
Village Development and Security Committee (JKKK) 147–148
voluntourism 150, 155, 157
WWF-Malaysia 199
Yayasan Emkay 199, 203, 205
see also Belum-Temengor forest reserve, Malaysia; Miso Walai Homestay, Sabah, Malaysia
Mao Zedong 45
May, T. 182
Mexico: community-based tourism 130
Mihalic, T. 141
Minca, C. 180
Miso Walai Homestay, Sabah, Malaysia 19, 141, *142*
background 142–143
'buy in' from local community and self-reliance 144–147, *146*
commercial viability and career paths 151–154
conservation 143–144
homestay variants 152, **152**
Kinabatangan Corridor of Life 148, 150, 154–155, 158, 160
KOPEL 20, 143, 144, 145, 146, 147–149, 150, *151*, 151–152, 153–156, 157–158
Lahad Datu incursion 158–159
local champions 147–149, *149*
Maandastay accommodation 152–153, 155
MESCOT CBT project 143, 145, 146, 147–148, 149, 154
NGOs 145, 154
ownership and pride in community 149–151
partnerships with specialist tour operators 156–158
philanthropic flows and local capacity 154–156
reforestation 155
responsible tourism 141–142
success factors 144–158
discussion and conclusion 158–160
Mitchell , J. 146, 154
Moeurn, V. *et al.* 145
monsoon tourism 13–14, 84–85
see also Hoi An community organic vegetable gardens
Moscardo, G. 178
MOTAC 155
mountain climbing 35
Murphree, M. 129

Nair, V. *et al.* 4, 5, 9
Nao, S. 69
natural disasters 14
nature-based solutions 14
Navrátil, J. *et al.* 77
Negros Island Certification Services Inc. (NICERT) 110
Nepal
mountain climbing 35
tourism policy 31–32, 33, 38
New Zealand
caves 42
Kaikoura Whale Watch 147, 153
Waitomo Glow-worm caves 42, 43
Ngo, T. *et al.* 138
NGOs (non-governmental organizations) 5, 19–20
Cambodia 164, 165
Kerala 97
Malaysia 145, 154, 203, 205
Timor Leste 180, 186
Vietnam 128

Nillson, P.A. 16
Nyapuane T. 32

Ooyama, Yoshiaki 65
Organization for Economic Co-operation and Development (OECD) 8, 9
Otsubo, S.T. 30
'overtourism' 11

Page, S.J. 1
Pereiro, X. 182
Perry, B. 182
Philippines
  Banaue rice terraces 113, 217
  Compostela Valley Province, Mindanao 52
  Department of Agriculture (DA) 107
  Department of Tourism (DOT) 106, 107
  Farm Tourism Act 112
  Republic Act 10816 107
  Technical Education and Skills Development Authority (TESDA) 111
  Tourism Master Plan (TMP) 106
  see also agritourism in the Philippines
Philippines Farm Tourism Industry Development Coordinating Council (PFTIDCC) 16, 107–108
poverty alleviation 50–52
Price, B. 81–82
Prideaux, B. 125
pro poor tourism (PPT) 55

Rawat, K. 8
reflexive practice 182
Reisinger, Y. 105
responsible rural tourism in Asia 1–3, 5–6, 11, 12, 20
  defined 2, 103
  framework 215, *216*
  requirements *vs.* characteristics 7, **7**
  responsible tourism 2
  rural tourism defined 8–11, **9**
  rurality defined 6–7
  sustainable tourism 1, 3–4, 5, 32, 88, 215
  see also agritourism in the Philippines; Asian forms of rural tourism; harmony between humans and nature
responsible tourism 2, 4–5, 20
  background 88–89
  defined 5, 32–33, 56, 87–88, 141, 173, 215
  see also Kerala: responsible tourism
Rich, S. *et al.* 106
Riessman, C.K. 182
Royal Society for the Preservation of Nature, Bhutan 36
Ruhanen, L. *et al.* 215
rule of law 18
rural areas
  defined 8–9, **9**
  tourism in 16–20, 214
rural tourism 16
  characteristics 213
  defined 2–3, 6–11, 213
  'family fun' 16–17
  managing stakeholders' challenges 180–181
  sense of place 13–14
  in small developing countries 179–180
  see also Asian forms of rural tourism; community-based tourism (CBT)
rurality defined 6–7
rural–urban migration 213
Ryan, C. 82

Salazar, N. 178, 182
Salazar, N.B. 117
Santilli, R. 117
Schipani, S. 164
Schroeder, K. 31–32, 33
Sebele, L.S. 117
sense of place 13–14, 82, 85, 104, 177, 217, 219
Sharpley, R. 32
Sikkim 12
Simkova, E. 8
Simpson, M.C. 10, 180
Sin, H.L. 180
Singh, S. 1
social capital 66, 69, 129
social empowerment 129
social equity 3
Social Exchange Theory 122–123, 181

Sonnino, R. 104
Sorenson, A. 16
South Africa: responsible tourism 33, 215
Sproule-Jones, M. 31–32
Sri Lanka: tea tourism 63–64
stakeholders 5, 10–11, 180–181, 197
Strauss, A. 200
Suansri, P. 10
Suharto 183
sustainable development 1
  defined 3, 88
  politics 3
  three pillars of sustainability 3, *4*, 5–6, *6*
  UN SDGs 125, 164
sustainable tourism 1, 3–4, 5, 32, 88, 215

Taylor, S.R. 129, 130
tea tourism
  China 64
  Sri Lanka 63–64
tea tourism in Japan 13, 61–62, *62*
  Higashi Sonogi (Nagasaki) **65**, 65–66, **70**, 71
  integrated rural tourism 63, **63**, **70**
  Kyoto Obobu Tea Farms 66, 67, 71
  Osawa (Shizuoka) 69, **70**, 71
  responsible tourism 4–6
  rural tourism 62, 63
  tea tourism 63–65
  Wazuka (Kyoto, Uji) 66–69, 70, **70**, 71
  discussion 69–71
  conclusion 72
Thailand
  rural areas **9**
  see also Belum-Temengor forest reserve, Malaysia
Timor Leste rural tourism 18, 177–178
  Balibo community 183, 188
  Balibo Fort Hotel (BFH) 184, 185, 187, 190
  Balibo House Trust (BHT) 183
  community development 186–190
  glocal ethnography 182
  Local Development Tourism 190–191, **191**
  local development tourism plan **191**
  managing stakeholders' challenges 177, 178, 179, 180–181, 182, **189**, 190
  National Strategic Development Plan (2011–2030) 182
  reflections on the project 185–191
  rural tourism in small developing countries 179–180
  study approach 181–183
  Timor Heritage Hotel (THH) 184, 185
  tourism capacity-building programme 183–185, 191–192
  tourism destination and nation building 181
  tourism narratives 178–179
  conclusion 191–192
Tolkach, D. 180
Tomocchi 66
tourism 87
  in rural areas 16–20, 214
  sustainable tourism 1, 3–4, 5, 32, 88, 215
  see also responsible tourism; rural tourism
tourism narratives 178–179
transformative tourism 105
Transparency Index 164
Turkey *see* Coruh Valley, Turkey
Turner, L. 181

'undertourism' 11
UNESCO World Heritage sites 79, 113, 131, 132, 218
United Kingdom: Department for International Development 51
United Nations (UN) 213
  Development Programme (UNDP) 17, 106, 116, 145, 169, 181
  Environment Programme (UNEP) 4
  General Assembly 1
  International Trade Centre 51
  Sustainable Development Goals (SDG) 125, 164
  World Food Program 168
  see also World Tourism Organization (UNWTO)
United States
  Federal Cave Resource Protection Act (1994) 42
  Government World Fact Report 162
University of the Philippines Asian Institute of Tourism (UP AIT) 107
urbanisation 213–214

values 18
Vietnam 75–76
   Youth News 79
   see also community-based tourism in Vietnam; Hoi An community organic vegetable gardens

Wangchuck, Jigme Singye 28, 30
Weaver, D.B. 16
wellness tourism 15
Willson, G.B. 77
World Bank 51, 162
   Governance Indicators Project (2017) 164
World Heritage Sites
   Banaue rice terraces, Philippines 113, 217
   Hoi An, Vietnam 14, 76, 79, 80, 132
   My Son, Vietnam 130–131
World Summit on Sustainable Development 5, 89, 103
World Tourism Organization (UNWTO) 4, 9, 87, 88, 90
   Global Code of Ethics 33, 197
World Travel and Tourism Council (WTTC) 89
WWF Norway 143, 145

Xu Honggang *et al.* 41
Xu Jing 90

Zapata, M.J. *et al.* 136
Zavadckyte, S. 67

For Product Safety Concerns and Information please contact our EU Authorised Representative:

Easy Access System Europe

Mustamäe tee 50

10621 Tallinn

Estonia

gpsr.requests@easproject.com